# EDITING IN THE MODERN CLASSROOM

*Editing in the Modern Classroom* is a research-based collection that defines the current state of technical editing pedagogy and plots a potential roadmap for its future. It examines current academic and professional editing practices, the global and corporate contexts of technical communication programs, and the role of new challenges such as content management in order to assess what should be expected from editing courses today and how instructors can best structure their courses to meet these expectations. It provides a research foundation to determine where changes are needed, and points to areas where additional research must be done to support further curricular and pedagogical innovations. *Editing in the Modern Classroom* challenges instructors to look deeper at the pedagogical aspects of what makes up an effective technical editing course at undergraduate and graduate levels and provides them with comprehensive and evidence-based resources to design and teach these courses.

**Suzan Flanagan** is the former managing editorial assistant for *Technical Communication Quarterly*. Before pursuing a PhD in rhetoric, writing, and professional communication, she worked as a freelance writer and editor. Her research focuses on technical and professional communication and the intersections of editorial processes, content strategy, and user experience.

**Michael J. Albers** is a professor at East Carolina University, where he teaches in the professional writing program. Before coming to ECU, he taught for eight years at the University of Memphis and worked for ten years as a technical communicator, writing software documentation and performing interface design. His research interests include methods for effective communication of complex information.

## ATTW Book Series in Technical and Professional Communication

Tharon Howard, Series Editor

*Editing in the Modern Classroom*
**Michael J. Albers and Suzan Flanagan**

*Creating Intelligent Content with Lightweight DITA*
**Carlos Evia**

*Involving the Audience: A Rhetorical Perspective on Using Social Media to Improve Websites*
**Lee-Ann Kastman Breuch**

*Rhetorical Work in Emergency Medical Services: Communicating in the Unpredictable Workplace*
**Elizabeth L. Angeli**

*Citizenship and Advocacy in Technical Communication*
**Godwin Y. Agboka and Natalia Mateeva**

*Communicating Project Management*
**Benjamin Lauren**

*Lean Technical Communication: Toward Sustainable Program Innovation*
**Meredith A. Johnson, W. Michele Simmons, and Patricia A. Sullivan**

*Scientific and Medical Communications: A Guide for Effective Practice*
**Scott A. Mogull**

For additional information on this series please visit www.routledge.com/ ATTW-Series-in-Technical-and-Professional-Communication/book-series/ ATTW, and for information on other Routledge titles visit www.routledge.com.

# EDITING IN THE MODERN CLASSROOM

*Edited by Suzan Flanagan and Michael J. Albers*

Routledge
Taylor & Francis Group

NEW YORK AND LONDON

First edition published 2019
by Routledge
52 Vanderbilt Avenue, New York, NY 10017

and by Routledge
2 Park Square, Milton Park, Abingdon, Oxon, OX14 4RN

*Routledge is an imprint of the Taylor & Francis Group, an informa business*

© 2019 Taylor & Francis

*Library of Congress Cataloging-in-Publication Data*
A catalog record has been requested for this book

ISBN: 978-0-8153-5445-1 (hbk)
ISBN: 978-0-8153-5446-8 (pbk)
ISBN: 978-1-351-13275-6 (ebk)

Typeset in Minion
by Swales & Willis Ltd, Exeter, Devon, UK

# CONTENTS

*Author Biographies*                                              *vii*
*Series Editor's Foreword*                                         *ix*
*Preface*                                                          *xi*

1   Editing in the Modern Classroom: An Overview                    1
    *Michael J. Albers and Suzan Flanagan*

2   The Current State of Technical Editing Research and
    the Open Questions                                             15
    *Suzan Flanagan*

3   "How Does That Make You Feel?" The Psychological
    Dimensions of Editorial Comments                               47
    *Ryan K. Boettger*

4   Imagination as Agency: Communities of Practice and
    Editing Pedagogy                                               66
    *Tracy Bridgeford*

5   Teaching Editing through a Feminist Theoretical Lens           91
    *Susan L. Popham*

6   Editing for Human–Information Interaction                     109
    *Michael J. Albers*

7   Concepts in Technical Editing Technologies: What's
    Important in Practice?                                    128
    *Clinton R. Lanier*

8   Editing for International Audiences: An Overview           146
    *Kirk St.Amant*

9   A Field-Wide View of Undergraduate and Graduate
    Editing Courses in Technical and Professional
    Communication Programs                                    171
    *Lisa Melonçon*

*Index*                                                       *192*

# AUTHOR BIOGRAPHIES

**Michael J. Albers** is a professor at East Carolina University, where he teaches in the professional writing program. Before earning his PhD, he worked for ten years as a technical communicator, writing software documentation and performing interface design. His research interests include designing for complex information and human–information interaction.

**Ryan K. Boettger** is an associate professor and assistant chair in the Department of Technical Communication at the University of North Texas. Previously, he worked as a technical editor for the Texas Army National Guard as well as the deputy editor-in-chief for *IEEE Transactions on Professional Communication*. He is the current editor of the Wiley-IEEE Press series in *Professional Engineering Communication*.

**Tracy Bridgeford** is a professor at the University of Nebraska at Omaha. In 2018, she published *Teaching Professional and Technical Communication*. In 2015, she co-edited *Academy–Industry Relationships* and, in 2014, she co-edited, *Sharing Our Intellectual Traces: Narrative Reflections from Administrators of PTS Programs*. She co-founded and edited *Programmatic Perspectives*.

**Suzan Flanagan** worked for *Technical Communication Quarterly* as a doctoral candidate. She researches technical and professional communication and the intersections of editorial processes, content strategy, and user experience. Her work has appeared in *Communication Design Quarterly, IEEE Proceedings of the ACM*, and the *International Journal of Sociotechnology and Knowledge Development*.

**Clinton R. Lanier** is a professor of Rhetoric and Professional Communication at New Mexico State University. Before earning his PhD, he worked as a technical writer for IBM, and as a Technical Editor for the US Army Research Laboratory at White Sands, NM.

**Lisa Melonçon** is an associate professor of Technical Communication at the University of South Florida where she also directs the undergraduate program. Her teaching and research focuses on the programmatic and professionalization dimensions of technical and professional communication, research methodologies, and the rhetoric of health and medicine.

**Susan L. Popham** is an assistant professor at Indiana University Southeast and teaches composition, business and professional writing, and magazine writing. She serves as a faculty advisor for a college magazine, for which her students research, write, and edit a wide variety of articles for publication in the magazine. She also has recently taken the position of editor for *Programmatic Perspectives*, the journal of the Council for Programs in Scientific and Technical Communication.

**Kirk St.Amant** is a professor and the Eunice C. Williamson Endowed Chair in Technical Communication at Louisiana Tech University (USA), and he is the Director of the University's Health and Medical Communication Center and its Usability Research Center. His research focuses on the globalization of online education and on health and medical communication for international audiences.

# SERIES EDITOR'S FOREWORD

Suzan Flanagan and Michael Albers' edited collection, *Editing in the Modern Classroom*, is the 11th in the ATTW Book Series in Technical and Professional Communication (TPC), and only our second edited collection. As its title suggests, this is a book about one of the most ubiquitous and pervasive subjects in TPC curricula—technical editing. Nearly every graduate, undergraduate, or certificate program in the nation offers courses in technical editing; in fact, as Lisa Melonçon points out in her examination of more than 600 degree programs in Chapter 9 of this collection, the technical editing course is generally considered a "core course" around which programs are built. Yet, in spite of its omnipresence in our field, surprisingly little ink has been shed in our current century that examines the subject in a critical way. As Flanagan observes in Chapter 2 of this collection, fewer than 20 empirical studies about technical editing have been published in our field's peer-reviewed journals since 2000, and those that have been published in this century display a surprisingly diverse definition of exactly what "technical editing" actually means and, perhaps more importantly, what we ought to be teaching in our technical editing classrooms in order to prepare our graduates for the modern workplace.

*Editing in the Modern Classroom* is an effort to bring technical editing into the 21st century. Like all the other books in the ATTW Book Series, this collection is solidly based on research and a comprehensive knowledge of the literature in the field. Flanagan and Albers have collected eight of the leading researchers and scholars in our field, and collectively they provide TPC students and practitioners with a multifaceted plunge into either the strategies and practices which working professionals use to practice technical editing, or into the effectiveness of how we teach technical editing. Flanagan and Albers' book is a useful antidote to those all-too-common situations where TPC instructors publish essays about "my favorite editing assignment" and yet fail to provide empirical evidence of the effectiveness of their pedagogies. Instead, through empirical studies of technical editing, the authors in this collection help us to understand the skills, strategies, and modern workplace technologies needed to prepare our students. The collection's authors ask us to examine the fundamentals upon which our modern technical editing courses are based and invite us to examine what we are teaching through modern-day lenses such as the psychological impact of commenting, editing as a communitarian practice, the user experience of editing, and the international dimensions of editing.

This book is an extremely timely and well-researched introduction to a pervasive and yet, ironically, overlooked topic in TPC. As such, it's a pleasure to have it in the ATTW Book Series in Technical and Professional Communication.

Dr. Tharon W. Howard
Editor, ATTW Book Series in
Technical and Professional Communication
October 13, 2018

# PREFACE

*Editing in the Modern Classroom* provides an examination of the curricular foundations that underlie the technical editing course. Using a research-based exploration, the authors of each chapter seek to determine where changes may be needed and where research is needed to support changes. The book works to open the conversation of exactly what should be expected from an editing course and how to structure it to achieve those expectations. Unless noted otherwise, the term *editing* refers to technical editing.

This book moves beyond the overly common practice of reporting the author's favorite aspect of editing, with teaching issues tacked onto the end, or describing "my favorite editing assignment." Both can be helpful, but neither forces instructors to look deeper at the pedagogical aspects of what makes up an effective technical editing course.

Little empirical research on the technical editing discipline has been published—a claim that applies both to pedagogy and to practice—which is a problem that the technical communication field needs to address (Boetteger, 2014, Chapter 3; Eaton, 2010; Flanagan, Chapter 2). The literature has many statements about the importance of editing (e.g., Corbin, Moell, & Boyd, 2002; Hayhoe, 2007), but then seems to ignore engaging in research on both how best to do it and how to best teach it. One goal of this book is to help start the conversation and to inspire more editing research.

This book strives to expose those deeper aspects.

- What are current editing practices and how can we prepare students to work within those practices? How is editing taught at both the undergraduate and graduate levels? What editing skills are considered essential to have mastered? How are instructors emphasizing the *technical* aspect of technical editing?

- How do past editing practices (good/bad/extinct) influence current teaching?
- Where does editing fit within the technical communication (TC) curriculum? How do/should undergraduate and graduate editing courses differ?
- How should a technical editing course be structured?
- How does the corporate shift toward globalization and content management change the pedagogy of the editing classroom? How should editing within a global context (nonnative English speakers as authors) be taught? Is there any difference?
- What do employers expect with respect to editing skills for their technical communicators and technical editors?
- How does the expansion of the technical editor's role (web design, visuals, content management, etc.) beyond "editing text" change the pedagogy of the editing classroom? How does editing compare/contrast with content management?

Some of the things the book is not:

- A collection of lessons learned while teaching editing.
- A collection of "my favorite assignment" in the editing course.
- A collection of chapters written to the student. In other words, this book is not designed to function as a textbook. Of course, many of the chapters are highly relevant and may be used to inspire discussion within a graduate editing course.

We hope this book in the short term helps to improve technical editing pedagogy and in the long term encourages more empirical research into both the practice and teaching of editing.

## References

Boettger, R. K. (2014). The technical communication editing test: Three studies on this assessment type. *Technical Communication*, *61*(4), 215–231.

Corbin, M., Moell, P., & Boyd, M. (2002). Technical editing as quality assurance: Adding value to content. *Technical Communication*, *49*(3), 286–300.

Eaton, A. (2010). Conducting research in technical editing. In A. J. Murphy (Ed.), *New perspectives on technical editing* (pp. 7–27). Amityville, NY: Baywood. doi:10.2190/NPOC2

Hayhoe, G. (2007). The future of technical writing and editing. *Technical Communication* *54*(2), 281–282.

# 1

# EDITING IN THE MODERN CLASSROOM

## An Overview

*Michael J. Albers and Suzan Flanagan*

## Introduction

*Editing in the Modern Classroom* is a research-based collection that both defines the current state of technical editing pedagogy and plots a potential roadmap for its future. The chapters are research-based and solidly grounded within the existing literature.

We see this introduction not as a place to answer questions about editing; the remainder of the book works toward that goal. But, instead, we see this introduction as a place to raise questions about editing, only some of which will be partially answered within the book. The rest will remain open research questions, which we hope this book will inspire researchers to begin to address.

We use *text* and *document* interchangeably throughout this chapter. The terms do not refer only to printed texts, but to any text, whether paper based, web based, or content management system (CMS) based, or other digitally created forms of text that a technical communicator may work with. At the discussion level here, the forms of text are all identical in that texts need to be created that effectively communicate their content; the actual presentation media used for that communication is more or less irrelevant here. We also use the term *editing* as shorthand for *technical editing*.

One of the aspects we emphasized when inviting chapters was that they should be research based. Overall, "the field of editing is remarkably light on empirical research," and some of that research is not directly "useful to a technical editing instructor, such as the pieces that examined whether copyediting improved the readability of medical manuscripts" through measurements using the Flesch-Kinkaid Readability Index, which is itself of questionable usefulness (A. Eaton, personal communication, May 15, 2017). (We agree that copyediting improves a text, but also acknowledge that a text can be grammatically perfect and still fail to communicate because of problems with content and/or presentation.)

Research into technical editing can be divided into two categories.

> *Methods of editing.* Research that looks at specific editing methods or techniques and measures their effectiveness, such as research that examined whether copyediting improved the readability of medical manuscripts (e.g., Roberts, Fletcher, & Fletcher, 1994).
>
> *Methods of teaching editing.* Research that looks at the pedagogical aspects of editing, such as how should students be taught to edit and what aspects of editing should be emphasized (e.g., Norman & Frederick, 2000).

However light the empirical research on practical aspects of editing, there has been essentially no empirical research on the technical editing course itself. What should a modern editing course contain? What should students learn? What is the relative importance of the preponderance of topics that could be included?

This book works to fill that gap and set the boundaries of what we already know about editing and editing pedagogy, and to state where the major holes exist—major holes that urgently need empirical research from technical communication researchers.

In an interesting twist, although there is a lack of empirical research, there is no dearth of pieces on the practical aspects of how to edit and on how editing is changing (e.g., Amare, 2009; Buttram, 2016; Schrank, 2013). But they tend to be opinion or anecdotal articles with no well-shaped research to back up their claims. We want to make clear that we separate empirical research on the editing course from either opinion or "how I teach" articles. Articles on "lessons learned," "my favorite assignment," or "what is unique about how I teach editing" are not uncommon and are how most questions about "how to teach editing" get answered. Unfortunately, the assignments and methods that work in one instructor's classroom rarely can be transferred directly into another.

Granted, many of those articles have gems of ideas and can help individual instructors inform their teaching. We have tried to incorporate some of those ideas into our editing courses. Some worked; some didn't. Some we try again next time; some we don't. However, the actual usefulness and impact of the techniques they describe are purely anecdotal. One instructor can say it worked wonderfully and everyone should use a technique, and we can shake our heads and say it was a dismal failure. Both are anecdotal opinions based on a single person's experience. What works well or fails with one classroom too often depends upon the dynamics of that one class and instructor. What made the idea or technique work for one instructor and not work for another—the underlying structural assignment issues—needs to receive our research attention. Empirical research studies can help us move beyond the individual class/instructor dynamic and uncover the pedagogical editing principles of what fundamentally works or doesn't work. Using those fundamentals, other people can make informed decisions to reshape their pedagogy.

Any teaching method or assignment is a complex interconnection of ideas. Making them work requires understanding both that complex web and what factors make them work or fail for students. Once that complex interconnection is acknowledged and is reasonably understood, then we have research that is generalizable to the editing course in general. Achieving this understanding is clearly not the result of a single study or the work of one scholar, but a longer-term research agenda pursued across the discipline. Unfortunately, more than just for the editing course, technical communication (TC) research in general seems to fail to try to verify the general applicability of the suggestions of "my favorite assignment." Hopefully, this book can help motivate researchers to undertake this work at all levels.

For example, consider a discussion among instructors about which order to teach the levels of edit. Many will argue for teaching copyediting first. One of us tends to teach that way, while the other tends to teach higher-level editing first. We both find that many of our students have trouble focusing on document-level issues of organization and content. Students are distracted by easy-to-fix lower-level issues. At times, we have had to explicitly state "no copyedit marks on the text" in order to force students to consider higher-level issues. The order-of-teaching issue is a fundamental structural issue that deserves research attention. This book doesn't address the order-of-teaching question, but it does highlight the extent of basic questions that remain open across the editing subdiscipline.

## Ways of Teaching

There are, of course, many different ways to teach an editing course. Our comments in this section are mostly written to an introductory editing course, but generally apply to an advanced course.

The introduction to technical communication and editing courses set the foundations of a student's understanding and eventual methods of editing practice. Additionally, the quality of the courses determines how well students adjust to editing in the workplace and how much time practitioners need to invest in training new employees. Workplace job structures are depreciating the editor's role, especially the comprehensive edit. The basic editorial functionality has not changed, but the percentage of the workday spent on it has greatly decreased. Rather than having dedicated editors, the job has morphed into peer writer/editor jobs with writers being responsible for editing each other's work. We can argue the validity of this workplace practice with respect to creating good texts, but regardless of our wishes, it reflects current reality. It remains a question of whether to try to incorporate this practice into the classroom or to teach as if the students are dedicated editors.

We must question—and empirically research—our current teaching methods. For instance, we should tackle some of the following research questions.

- How are students being taught to edit?
- Does the course include client projects? Does the course include strategies for managing clients?
- Are students being taught to edit texts from nonnative English speakers?
- Does the course include editing text that will undergo translation and localization? As St.Amant points out in Chapter 8, these texts require specific changes that often make no sense to a person seeing the text through a single-language lens.
- What do undergraduate and graduate editing courses look like? Does a student who takes them in different semesters find distinctively different courses or is the graduate course more or less a repeat of the undergraduate course? (Yes, we acknowledge that most technical and professional communication graduate programs do not require the students to have had an undergraduate editing course.)
- Is our definition of the editing course too narrow and should it include the broader functions a practicing editor is called upon to perform: document management, publishing management, coordinating revisions and reviews, etc.?

These initial questions point to additional research questions about the editing course, some of which Bridgeford and Melonçon explore in Chapter 4 and Chapter 9, respectively.

- Should we continue to teach print-based editing techniques, such as hard copy markup?
- Are the movement of text to content management systems and other online writing methods being taught? Effective editing here is not a direct transfer from paper editing.
- How does editing fit within collaborative writing? How are these practices being taught?
- Does the course include editing situations in which the editor doesn't fully comprehend the material? There is no guarantee that the writer produced a coherent text that contains the proper content in the proper order. Even here an editor needs to make effective comments.
- How can students demonstrate their writing and editing skills to potential employers? How can programs provide opportunities for students to demonstrate these skills?
- How do we define *editing* across TC programs?

And then there is the question of learning and knowledge transfer. Technical editing discussions tend to focus on what we are teaching. The follow-on question of what students are actually learning is rarely addressed. If it is addressed, the answers tend to be anecdotes describing how animated and engaged students are in the class—which is definitely a good thing. Unfortunately, it is one thing to include a topic in a course, even one that gets the student engaged, and another for a student to be able to coherently discuss the topic after they complete the course. And something else again for them to know to apply it to a different situation.

Let's look at three different ways an editing course could be taught. We realize that many courses are actually a mashup of these different ways.

### Copyedit and Grammar Course

When students start an editing course, many assume they will be learning copyediting and risk being transformed into that person who gleefully covers pages in red ink.

They also think "editing" starts and ends with the copyedit. This, of course, is a holdover from the deeply ingrained prior knowledge—driven deep by high school and freshman composition—that editing equals proofreading just before submitting a paper. The use of "edit your work," "copyedit your work," and "proofread your work" as equivalent phrases does everyone a disservice and requires the student to unlearn material.

A copyedit-based course comes with a strong focus on grammar. In this format, many lectures may be devoted to discussing sentence structure and grammar, and the student probably does many worksheets, each of which addresses one type of grammar problem, such as "edit these sentences for proper tense." The learning transfer from such worksheets is questionable (Crovitz & Devereaux, 2017; NCTE, 2008). A copyedit-based course typically has an overly academic editing focus with test questions such as "take this information and properly mark it for APA format."

We also see deeper curricular issues. We agree that many students currently lack and can benefit from a deeper understanding of English grammar, but we question the big-picture pedagogy that transforms an editing course into a grammar course. A strict copyediting course teaches editing equals "make grammatically correct." Of course, after editing, a document should have correct grammar, but if the overall document structure or content fails to communicate or fails to meet the users' needs, then the end result is a worthless, but grammatically perfect document. In other words, the company, the authors, and the editors have wasted their time and money.

We could be guilty of creating a strawman argument in the previous paragraph, but we fear it is not as strawman as many of us wish it were. We've had too many conversations at Society for Technical Communication (STC) conferences with technical writers who consider it very important to go over a text multiple times to ensure there are positively no grammar errors (the content came from marketing or an engineer and the writer doesn't worry about the accuracy—they just format what they are given—but the content must have perfectly written sentences). They admit lacking knowledge to judge the content, so they focus on grammar and layout. Worse, they don't seem to care how, why, or even if their documents are used. Many of these writers see their job as taking text from programmers/engineers, cleaning up the grammar, and properly formatting it . . . end of task. Sadly, they have redefined technical communication as being a copyeditor/layout person.

From an industry perspective, someone hired as a full-time copyeditor is viewed as a clerical person. Why would we, tasked with educating people to receive a college degree, want to prepare our students for clerical jobs with minimal advancement potential, rather than for professional jobs that require symbolic-analytic knowledge work (Dicks, 2010; Johnson-Eilola, 2004)? When this happens at the graduate level, we see it as highly problematical.

### Comprehensive Editing Course

The most important aspect—and most difficult to learn—of technical editing is learning how to evaluate a text at a level higher than a sentence level. In other words, how to perform a comprehensive edit. Learning to perform comprehensive editing is learning how to evaluate a text's structure and analyze the ability of that structure to effectively communicate with the document's audiences within their context. This ability is difficult to master because (1) most students are not technically knowledgeable in the subject or context, and (2) a document has *audiences*, plural noun, not *an audience*, singular noun.

One scenario to ask students to consider early in a class:

> There is a single report on one single topic . . . One author's take is "I'm writing technical information" and another author's take is "I'm communicating technical information." Do they have the same view of the writing process? Why?

Another question to ask students, which directly ties to editing pedagogy:

> What makes good business or technical writing? Conversely, what makes bad business or technical writing? Does good technical writing relate closely to any of the following ideas?
>
> - The text is grammatically correct.
> - The text contains all the required information. (Define what "all the required information" means.)
> - All the information is correct. (Define what "correct information" is.)
> - The text uses graphics.

Student responses to the first question should focus on issues of audience. Any text worth writing has content that needs to be communicated to an audience. The editor works to verify that communication should occur. The first writer, who is "writing technical information," dumps all the information onto a page with the view that it is the reader's job to figure it out.

Many students' initial response to the second question is that "if the spelling and grammar are bad, you lose credibility." A true statement, but that's not the question being asked here. In the professional world, you can assume the spelling and grammar are correct and the document is technically correct. It's not the obvious things that make a document bad. But can doing all the obvious things still result in a bad document?

The question a technical communication student needs to consider is not whether good technical writing contains those things, but the opposite . . . Can someone say, "It is technically correct, has good grammar, and has graphics; therefore, it is good technical writing." Is an editor's primary job to ensure the text meets these criteria? Part of the job, yes; primary job, no! Our response, and one that most students eventually realize, is that these things are necessary, but not remotely sufficient to make a good text. Thus, one goal of an editing course is to provide a skill set that can be used to analyze a document, make the judgment calls for how to make sure the document communicates information, and show how to communicate those judgments to the author. Editing is not about creating a grammatically perfect text, but about creating a text that effectively communicates its information to its audience.

Actually, we consider both questions to be different aspects of the same question. An editor must be able to analyze the text with respect to both desired content and audience. Editors who view their job as creating a grammatically correct and format-correct document are doing less than half of their job. Addressing these questions within an editing course requires an approach to editing that moves well beyond copyediting. Rather, such contexts require a guided form of comprehensive editing that can involve actions from changing certain words, to reorganizing sections, to requesting additional content, or to working with the authors to revise passages of text. Unless a student has gained these skills, they will be unable to effectively perform a comprehensive edit.

Perhaps just as important, and not something that can be easily taught within the limitations of any writing course, is editing not just the text before you, but editing to ensure the text belongs in that document. Too many technical documents suffer from either too much extraneous information or too little relevant information. An editor needs to be able to examine the text and comment on both. Looking beyond the words on the page to words that are not there or words that should not be there involves a steep learning curve. But the ability to see beyond the page is a vital skill an editor must develop. A single editing course will probably not be able to teach this skill at more than a trivial level, but it is important to make students realize it is a major element of comprehensive editing.

For example, Albers once held a writing job where the editors controlled the libraries. Writers were assigned individual books within the product library. The company made mainframe software products, so the libraries for each product were huge. From a writer's view, Albers may have wanted to talk about X in his book. It made sense to him to talk about it. However, if he did try to write about it, he'd probably start from scratch explaining the topic and work up to what he needed with respect to his book. In the bigger picture, the editor said no because all of that was already in a different book. Albers would be duplicating effort, making maintenance more difficult, and making his book harder to read. Perhaps this is an extreme example of editors having to draw the line on content, but it is one that students need to understand.

Seeing any work within a bigger picture is not something students have experience with. They have spent essentially their entire education career writing one-off documents (write—turn in—done) and being told to "do your own work." Mention group projects and complaints immediately begin about how it would be faster to just do it alone, or the project gets divided into separate pieces with each student working individually. Thus, students are geared to thinking in terms of writing a text, not in terms of analyzing or editing a text. Few have received explicit instruction on how to edit or comment.

As a result, it's amazingly difficult to convince some students to *edit* a text, rather than rewrite it. Maybe the students can say it better than the author, and it is definitely easier to just do the rewrite rather than try to make a bunch of comments, but the students need to learn that an editor's job is to comment, not to rewrite. In Albers and Marsella's (2011) study on editorial comment structure, two students made no comments; they jumped directly to rewriting the text. The study was conducted about two-thirds of the way through the semester of an introductory editing course and some students still failed to grasp the idea of commenting.

In addition to teaching students how to perform an editorial analysis of texts, we need to teach students how to construct comments. Teaching students to edit within the corporate environment must also include having them overcome the fear of upsetting people by making suggestions to a text. Even beyond the basics of comment construction, such as never say, "hey, orc face, fix this or else," students need to gain confidence at stating what needs to be done. Many students are very good at analyzing and commenting on texts supplied by the instructor. The students show a solid ability to make constructive comments when the text has no real author. But later in the course, students have to do a comprehensive edit of each other's work. Suddenly, comments become scarce and are little more than copyedit level—the bad habits that instructors tried to quash in freshman composition peer-review workshops reemerge: Students give the text a quick read through, declare "the text is great and ready to turn in," and discuss weekend activities.

### Specialized Editing Course

Most TC programs only have one undergraduate editing course, for a range of valid programmatic reasons. And perhaps a two-course sequence in editing is not really needed—if the program incorporates editing as a significant element of other courses. But there is also a need to consider courses focused on several specialized aspects of editing—we're using *specialized editing* for lack of a better term. This specialized editing course would be any advanced course that has a specific focus: web, usability, technology, international, management, or a different lens (such as social constructionism, feminism, and postmodernism).

Editors need to learn how to work with and manage people within an increasingly technical environment. According to D. Kain (personal communication, May 15, 2017),

> editors often have management responsibilities (Meyer, 2009; Petelin, 2002); they need to understand budgets, ROIs, and people and project management (Eaton, Brewer, Portewig, Davidson, & Portewig, 2008; Hackos, 2006; Wilde, Corbin, Jenkins, & Rouiller, 2006) in addition to being concerned with grammar, style, document design, and information architecture.

With the management responsibilities comes the requirement that editors need to be able to talk with the writers, the subject matter experts, and management. In addition, editors, excluding the clerical-level copyeditors, should sit high in the hierarchy. (That said, the corporate world has moved away from a hierarchical structure and into a matrix-based structure.) Editors should handle the product library with individual writers assigned to specific books within that library. Obviously, the idea of books and library changes with online documentation and content management, but the entire product information structure still requires a single overseer.

### Different Theoretical Lens

As Popham explains in Chapter 5, there is little scholarship on editing with any specific theoretical lens. Much like the traditional view of technical communication, "editing is often viewed as an objective, neutral, pragmatic work, devoid of pesky or troubling theoretical approaches" (hmmm, is "must have perfect grammar" a theoretical approach?). "Extending this perspective to the teaching of editing, editing as a course subject continues to be viewed in much the same light—straightforward, pragmatic, neutral, and objective" (Popham, personal communication, May 15, 2017).

Like Popham, we find it interesting that rhetorical approaches to editing are rarely specified even though some scholars have embraced rhetorical approaches to grammar (e.g., Connatser, 2004; Kolln & Gray, 2017; Micchiche, 2013).

> Rhetoric pervades technical communication pedagogy; rhetoric, the study and application of audience, textual, and purposeful strategies for persuading audiences through textual (and contemporarily, extratextual) means is at the heart of any editing goal—to help the author effectively persuade one's audience through the best textual means. (Popham, personal communication, May 15, 2017)

A student (or instructor) with a basic rhetoric background will immediately point out that no rhetorical situation is neutral and objective, and then proceed to editing as if it is. At the copyedit level, editing more or less is neutral. English has a set of grammar rules that must be followed. A corporation has a style guide which must be followed. Of course, how closely they are followed and which specific instances are given a pass can call into question the "neutral and objective" aspect. Once we get beyond copyediting, the higher levels of edit are ripe with rhetorical decisions about improving communication and are not remotely neutral and objective.

Beyond issues of the rhetoric of editing, there are the various theoretical lenses that could be used to examine how to edit a text. In Chapter 5, Popham looks at how feminist theory fits within editing. Various cultural lenses—of the type TC studies as cultural rhetorics rather than editing for international audiences—could also be applied. Perhaps one of the reasons this type of lens is not applied more is that editing is but a step in creating a document. Most of the studies that analyze texts through different theoretical lenses tend to examine a completed document and not the process of creating the document. The analysis conclusion typically includes a call to be aware of whatever is being analyzed. A detailed examination of how editors contribute to or inhibit the different views would be interesting.

## International Editing

Increasingly, technical materials are being sent to a world-wide audience. This distribution practice requires at a minimum that the text be translated and may require localization or globalization. Or it may require the text to conform to more standard, or global, English for use with international audiences (St.Amant, 2002).

The editor is gatekeeper for ensuring texts conform to proper standards. The text and the layout need to be edited to ensure the text conforms to these international and cultural issues as well as to any legal and accessibility requirements. As a stand-alone text, the document may be fine, but if it will be translated or if the text (included after translation) must fit within specific areas (e.g., graphic call outs), the editor must point out any required changes.

Editors need to understand the management issues associated with translation and localization. In addition, they need to understand the issues nonnative English readers have with reading technical texts. The technical communication literature contains many articles looking at these issues, but they focus on writing. Few specifically address the editing issues. We also need work on how to motivate students on these types of assignments when most of them are not fluent in a second language and probably do not understand the global nature of modern corporations.

Even for texts that only require the use of U.S. English, there are still cultural issues. The reading levels need to conform to the document's intended audiences and the overall design must fit their perceptions of the document.

## Tools and Technology of Editing

The tools and technology of writing and editing within the corporate environment are changing rapidly and resulting in a change of the editorial role. The person sitting at a desk with a dedicated job description and title of "editor" is highly endangered. Although the editor is disappearing from the corporate workspace, the editing still happens. Each writer becomes a writer/editor working with the other members of their group. Or they work as a writer on one project and are charged with editing another. The job of the corporate editor is changing—but not disappearing, as some claim—to accommodate changing work environments and technologies.

Technical editing pedagogy needs to change to reflect these new realities of the lack of dedicated editors and the addition of tools for controlling and managing texts. In the end, it is the editor—however defined—that constructs the final documents and ensures they fit the audience needs. We need research on how to efficiently perform this process within the new environment. The current methods need to be reexamined since they tend to be old-school print-based methods.

Along with these job and technology changes, perhaps we need a more integrated TC curriculum where issues such as editing, tools, research and such appear at different levels of detail across all courses. Well outside of the scope of this chapter is the problematic issue that editing (and too many other TC topics) is taught using a silo method of teaching. This course teaches X; connecting X to the rest of the program is left as an exercise for the student.

## Nontext Editing

An editing course could also examine the nonword aspects of editing—editing that focuses on features beyond the text, such the visuals, document usability, or information architecture. In some instances, writers are being asked to create the entire document, including the layouts. Editing has shifted to include the nontext parts of a document.

In a text-based editing course, instructors may find that asking students to edit elements other than words is helpful for teaching comprehensive editing.

For example, when the first day of class coincided with a solar eclipse, Flanagan showed her students online instructions and how-to videos and provided materials for making two types of solar viewers from cardboard boxes and paper plates. Even before the students attempted to make their own viewers, they found problems with the directions, explanations, graphics, and safety warnings (or lack of). No one discussed grammar problems. Instead, they asked questions about logic (e.g., why did the eyehole placements differ between models? Why was tin foil needed?) and noted graphic design problems (e.g., a yellow image on a white background). Problems with missing information, organization, audience awareness, and usability became apparent when students tried to build their viewers (e.g., incomplete material lists, assembly order, difficulty level).

Hands-on nontext editing activities such as this one may help students view editing in terms of communication problems and help them focus on big-picture comprehensive editing issues; however, if a nontext-based editing course were offered, we can see that the course would integrate both the creation and editing aspects. Unfortunately, the editing would assuredly be subservient to the creation aspects. It would be similar to making an argument that a program does not need an editing course because editing is covered in the introductory TC course. Yes, it has one unit, but is that adequate?

Some aspects such a course could cover are as follows.

- Edit the visuals. Do the visuals work for the text? This process includes both editing the textual aspects of the visual and ensuring the visual itself is appropriate for the text. The visual may be very well constructed, but doesn't aid in communicating the text's message.
- Determine whether a text should have more visuals. Too often the existing visuals may be well formed and they get edited well. But are they sufficient? Does the text require more?
- Edit for usability. Does the information architecture of the text (both web and print) work? Can people find and refind information?

## Book Takeaways

Many of these takeaways are worded for the editing course, since that is the topic of this book, but they really apply across the entire TC curriculum at both the undergraduate and graduate level.

- We need more editing research at both the practical editing and editing pedagogy level. The suggestions of "my favorite assignment" need to be formally researched.
- We need to define what the editing course should contain and how to best teach that material.

- We need to understand the needs, requirements, and expectations of the changing corporate world. How does the curriculum need to change to reflect those changes?

## References

Albers, M. J., & Marsella, J. F. (2011). An analysis of student comments in comprehensive editing. *Technical Communication, 58*(1), 52–67.

Amare, N. (2009). The technical editor as new media author: How CMSs affect editorial authority. In R. Spilka (Ed.), *Digital literacy for technical communication: 21st century theory and practice* (pp. 181–199). New York, NY: Routledge.

Buttram, S. M. (2016). Changing the rules in an old, established game: Editing in engineering-based government agencies. *Intercom, 63*(7), 10–12.

Connatser, B. (2004). Reconsidering some prescriptive rules of grammar and composition. *Technical Communication, 51*(2), 264–275.

Crovitz, D., & Devereaux, M. (2017). *Grammar to get things done: A practical guide for teachers anchored in real-world usage.* New York, NY: Routledge and National Council for Teachers of English.

Dicks, R. S. (2010). The effects of digital literacy on the nature of technical communication work. In R. Spilka (Ed.), *Digital literacy for technical communication: 21st century theory and practice* (pp. 51–81). New York, NY: Routledge.

Eaton, A., Brewer, P. E., Portewig, T. C., Davidson, C. R., & Portewig, C. C. (2008). Comparing cultural perceptions of editing from the author's point of view. *Technical Communication, 55*(2), 140–166.

Frost, E. (2014). Apparent feminist pedagogies. *Programmatic Perspectives, 6*(1), 110–131.

Hackos, J. T. (2006), *Information development: Managing documentation projects, portfolio, and people.* New York, NY: John Wiley and Sons.

Johnson-Eilola, J. (2004). Relocating the value of work: Technical communication in a post-industrial age. In J. Johnson-Eilola & S. A. Selber (Eds), *Central works in technical communication* (pp. 175–194). New York, NY: Oxford University Press.

Kolln, M., & Gray, L. (2017). *Rhetorical grammar: Grammatical choices, rhetorical effects* (8th ed.). Boston, MA: Pearson.

Meyer, J. R. (2009). Effect of primary goal on secondary goal importance and message plan acceptability. *Communication Studies, 60*(5), 509–525. doi:10.1080/105 10970903260343

Micchiche, L. R. (2013). Making a case for rhetorical grammar. In S. N. Bernstein (Ed.), *Teaching developmental writing* (4th ed.; pp. 220–238). Boston, MA: Bedford / St. Martin's.

National Council of Teachers of English. (2008, Nov. 5). Beyond grammar drills: How language works in learning to write. *Council Chronical Online.* Retrieved from http://www.ncte.org/magazine/archives/125935

Norman, R., & Frederick, R. A. (2000). Integrating technical editing students into a multidisciplinary engineering project. *Technical Communication Quarterly, 9*(2), 163–189. doi:10.1080/10572250009364692

Petelin, R. (2002). Managing organizational writing to enhance corporate credibility. *Journal of Communication Management, 7*(2), 172–180.

Roberts, J. C., Fletcher, R. H., & Fletcher, S. W. (1994). Effects of peer review and editing on the readability of articles published in *Annals of Internal Medicine*. *JAMA, 272*, 119–121.

Schrank, K. (2013). Using editing checklists for more efficient editing. *AMWA Journal, 28*(4), 164–166.

St.Amant, K. (2002). When cultures and computers collide: Rethinking computer-mediated communication according to international and intercultural communication expectations. *Journal of Business and Technical Communication, 16*(2), 196–214.

Wilde, E., Corbin, M., Jenkins, J., & Rouiller, S. (2006). Defining a quality system: Nine characteristics of quality and the editing for quality process. *Technical Communication, 53*(4), 439–446.

# 2

# THE CURRENT STATE OF TECHNICAL EDITING RESEARCH AND THE OPEN QUESTIONS

*Suzan Flanagan*

## Chapter Takeaways

- The term *technical editing* does not have a well-established definition.
- Technical editing pedagogy and curricula are based largely on personal experience and lore.
- Fewer than 20 empirical studies on technical editing have been published in peer-reviewed technical communication journals since 2000.
- Most of the empirical research on technical editing focuses on editorial comments, electronic editing, editing tests, and quality control.
- Electronic editing tools are well established in the workplace, and editing work is shifting to cloud-based collaboration.

## Introduction

Communication tools and practices have changed considerably since the technical communication field began. Long gone are the once-ubiquitous typewriters, word processors that displayed only a few lines of text, command-driven computers with floppy drives, bulky monitors with green, amber, or white text, and dot matrix printers that screeched. Although obsolete, those technologies—like their replacements—allowed technical communicators to develop, display, and distribute information.

Demands for information have increased with the rise of two-way communication channels such as the internet and social media. Facilitated by technology, information has become a commodity, which, without professional editorial oversight, often lacks quality and credibility. Computer algorithms and artificial intelligence have not yet mastered the complexity of language.

For the foreseeable future, editors will be needed to optimize the communication of—and interactions with—information. However, as technologies evolve, editors must update their skills and approaches to editing.

Editing is a core skill for technical communicators, yet the amount of literature on the topic doesn't seem to reflect the critical role technical editors play in shaping information into usable and useful content for audiences. The nature of the technical editing literature as a whole is problematic in that much of the literature published prior to 2000 consists of anecdotal commentaries, lore-based tutorials, and informal research, all of which typically lack the theoretical grounding and rigorous methodology of empirical research (Eaton, 2010). With limited empirical research, it is difficult for the field to advance. We need empirical evidence to do the following:

- make editorial work visible
- create a solid business case for editorial work and editors' expertise
- confirm or refute that current best practices are indeed best practices
- establish best practices for emerging publication models and technologies
- test employers' equivalency assumptions that writers and editors are interchangeable
- explore the limitations of AI as an editorial tool in global, dynamic environments
- verify whether technical editing curricula align with workplace practices.

In reviewing the technical editing literature from 2000 onward, my goals are to (1) summarize the findings of previous empirical research studies; (2) identify areas where further empirical research is needed to help validate or refute lore-based, anecdotal knowledge or to test and/or extend the findings of previous empirical studies; (3) pinpoint knowledge gaps; (4) connect the research to editing pedagogy; and (5) compile open research questions with the potential to improve the practice and teaching of technical editing. I use MacNealy's (1999, pp. 40–41) "essential characteristics of empirical research" as criteria for identifying empirical studies:

- "An empirical study is planned in advance of the data collection."
- "The data are collected systematically."
- "The method of data collection produces a body of evidence that can be examined by others."

## Scope

To locate relevant literature, I used the search terms "technical editing," "technical editor," and "editing" AND "technical communication." Although extensive, my search of the literature was not comprehensive and did not include editing research from fields outside technical communication. Nor did my search include materials published prior to 1980, materials that required interlibrary loans, or materials available locally on microfilm only, such as *The Technical Writing Teacher*, which was replaced by *Technical Communication Quarterly* in 1992. Also excluded were online resources, such as the editing section of the STC Technical Communication Body of Knowledge (Fitzgerald, Stanley, & Lindstrom, 2017) and the EServer Technical Communication Library: An Annotated Bibliography of Technical, Scientific, and Professional Communication (https://tc.eserver.org).

This literature review focuses on peer-reviewed articles about technical editing that have been published primarily in the 21st century by major technical communication journals (i.e., *IEEE Transactions on Professional Communication, Journal of Business and Technical Communication, Journal of Technical Writing and Communication, Technical Communication*, and *Technical Communication Quarterly*). In keeping with the empirical orientation of this book, I privilege articles that report empirical research but do not dismiss the value of other forms of knowledge-making. Despite my emphasis on empirical research, I have excluded conference proceedings due to space constraints and authors' tendencies to later publish fuller accounts of the research in peer-reviewed journals or books. I do, however, briefly discuss the most recent editions of several technical editing books.

## Content Analysis

Before I delve into the 21st-century technical editing literature, I will point out a few notable landmarks among the earlier literature that capture the conversations that have occurred since the 1980s. Looking back to the era when command-driven, floppy-disk-based personal computers emerged and print-based publication models were the norm, we find in the final quarter of 1981 a special issue of *Technical Communication* devoted to technical editing. H. Lee Shimberg (1981b) guest-edited the issue, which featured articles by Lola Zook, Mary Fran Buehler, Don Bush, Alberta Cox, Harold Osborne, Eva Dukes, and Shimberg (1981a). Topics ranged from editing definitions and levels of edit to author–editor relationships and the value of editors and editing. Today's readers would expect a special issue to cover topics such as editing content for mobile devices or other digital publication models; editing content for usability, accessibility, or global audiences; and editing content with automated and cloud-based tools. Discussions of editing practices as they intersect with project management, information architecture, or content strategy would be valuable too.

Since the 1980s—and perhaps earlier—much of the literature on technical editing appeared in *Technical Communication* and *Intercom*, both of which are publications of the Society for Technical Communication (STC). (As a magazine for a professional society, *Intercom* is not peer reviewed; however, since the publication contains multiple two- to six-page articles on technical editing that would be useful in the classroom, I briefly discuss them.) In my search of the literature, I found approximately three times as many editing-related pieces published in *Technical Communication* compared with the other major technical communication journals. *Intercom* has published a similar number; however, I did not have access to issues prior to 2000. Figure 2.1 shows the approximate number of editing-related articles appearing in each publication by time period; the actual numbers may be higher.

Notably, within the category "Other," the *American Medical Writers Association Journal* published four editing-related articles, which exceeds the number of articles published by any of the peer-reviewed technical communication journals (except *Technical Communication*) during the same period. Two articles were published by *JAMA*. The rest were published in various publications: *British Journal of Educational Technology, Business Communication Quarterly, Composition Studies, Convergence, International Journal of Sociotechnology and Knowledge Development, Issues in Writing, Journal of Educational Technology Systems, Journal of English for Academic Purposes, Research in the Teaching of English, Triple C*, and *Written Communication*. When empirical-focused journals such as *Written Communication* publish only one empirical-based editing article within the last decade, we can infer that little research on editing is being done.

Yet a demand for such research seems to exist. Combined, *Technical Communication* and *Intercom* have published more than 50 editing-focused

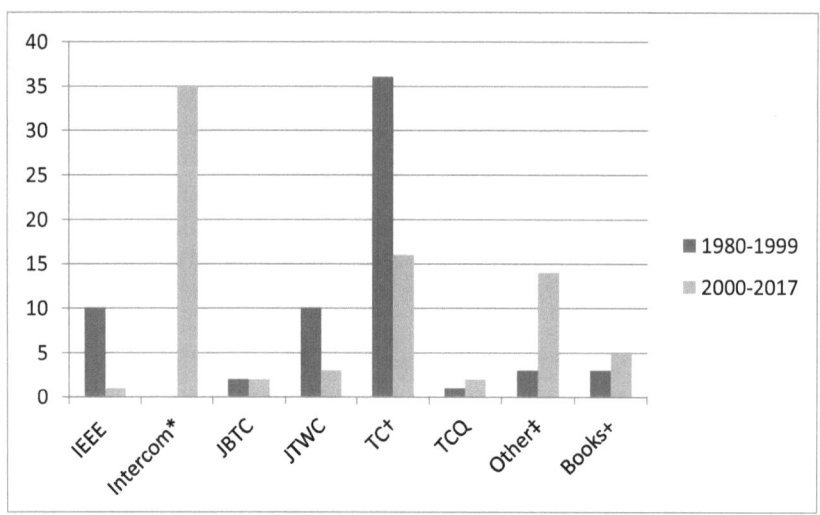

**FIGURE 2.1** Editing-related articles by publication and time period.

*Data prior to 2000 were not analyzed.
†The 1980–1999 figure includes seven columnist pieces; the 2000–2017 figure includes two editorials.
‡Other includes 16 different journals covering disciplines such as medicine, technology, and English-related fields.
+The number of books is underrepresented particularly in the 1980–1999 period because prior editions and out-of-print books were not included in the review.

articles in the last ten years. This relative abundance of editing articles may reflect the nature of the STC's membership and the publications' editorial missions. Both publications serve academics' and industry practitioners' needs. Practitioners value how-to information that helps them solve day-to-day communication problems and provides them with best practices for performing core tasks such as editing. Titles such as "How to Edit the Passive Writer's Work" (Mann, 1985), "The Editor's Nightmare: Formatting Lists within a Text" (Plunka, 1988), and "Editing Modular Documentation: Some Best Practices" (Strimling & Corbin, 2009) would appeal to practicing editors yet also prove useful to academics who teach editing and are looking for current information to supplement—or to fill gaps in—textbook content.

The *Intercom* articles deliver practical, easy-to-understand advice based on the authors' experiences, best practices, and a few outside sources. Among the other technical-editing topics discussed are

- commenting strategies (e.g., Cryer, 2012)
- editor roles and relationships (e.g., Cundiff, 2017; Hallmark & Sowards, 2009; Henkin, 2017; Moell et al., 2012; Murphy, 2015)
- editing skills (e.g., Brown-Hoekstra, 2017; Collins, 2001; Hart, 2003; Lemanski, 2012; McNeill, 2001; Thomas, 2009)
- plain language (e.g., Gillenwater, 2017)
- second-language authors and global audiences (e.g., Ketelaar, 2001; White, 2012)
- style sheets (e.g., Frick & Frick, 2009; Hallmark, 2009)
- technology (e.g., Cantella & Corbin, 2012; Glick, 2010; Radella, 2000; Seagren & Gardash, 2012).

In addition, columnists Don Bush (The Friendly Editor, *Technical Communication* and *Intercom*) and Geoffrey Hart (Effective On-Screen Editing, *Intercom*) offer pearls of technical-editing wisdom on a variety of topics:

- commenting strategies (Hart, 2002c)
- content elements and genres (Bush, 1994b, 2001b, 2002b; Hart, 2002b)
- editing skills (Bush, 1993a, 1994a, 2000)
- editors' value (Bush, 1991)
- grammar (Bush, 1994c)
- language (Bush, 1993b)
- careers (Bush, 2001a, 2001c, 2002a)
- style sheets (Hart, 2002a)
- technology (Bush, 1992; Hart, 2000, 2001).

Both lists include evergreen topics—timeless topics, such as style sheets, that may have been refreshed to reflect current practices—and contemporary issues, such as cultural trends, shifts in business practices, and technological advancements. *Intercom* is published more frequently than the peer-reviewed journals and usually plans its editorial calendar for the year. In contrast, the peer-reviewed journals typically plan one or two special issues per year.

While it's not uncommon for publications to avoid repeating topics within a specified time period (e.g., *IEEE Transactions on Professional Communication* waits five years until repeating a special issue topic), 17 years elapsed between *Technical Communication*'s special issue on editing and the next cluster of technical editing articles. (I use the term cluster to describe occurrences of more than two articles on the same topic within the same publication issue when there are no explicit indications of a special issue, special section, or designated theme.) In 1998, the first-quarter issue of *IEEE Transactions on Professional Communication* included six articles on editing and one article on an unrelated topic; the former topics included levels of edit; editing visual media, websites, and intranet content; international editing, style guides, and ethics. Fourteen years later, the September 2012 issue of *Intercom* focused on editing; the articles discussed editing tips, editing skills, international editing, editor roles, and editing specialized subject matter. Another five years later, the July/August 2017 issue of *Intercom* featured four editing-related pieces that covered reviewing and editing, plain language, and editor roles.

Nearly 40 years after the *Technical Communication* special issue, we're still grappling with many of the same topics discussed in that issue: editing definitions, levels of edit (e.g., copyediting to substantive editing), style, author–editor relationships, career opportunities and challenges, the size of the field, and the perceived value of editors and editing.

## Editing Definitions

The convention of defining key terms is an ingrained practice for technical communicators, yet the term *technical editing*—or even *editing* itself—does not have a well-established definition. Technical communicators seem to agree that editing is a process, but the process may be defined in terms of technology, rhetoric, actors, activities, and/or disciplines—intersecting categories that I have used below to compare nuanced meanings.

### Technology-Based Definitions

Some technical communicators (e.g., Coggin & Porter, 1993; Rude & Eaton, 2011) first define editing and then complicate the definition by appending the term technical. Others (e.g., Amare, Nowlin, & Weber, 2011) begin by situating technical editing within the field of technical communication and by defining technical editing as a form of editing. Most develop extended definitions over paragraphs, if not pages; the definitions below provide snapshots of longer ones.

| Coggin & Porter (1993) | **Editing**: "the process of modifying text to prepare it for publication; may involve several types of edits from substantive revision to minor proofreading" (p. 233). |
| | **Technical editing**: a type of editing that involves technical and scientific content, technical accuracy, and consultation with subject-matter experts. |
| Rude & Eaton (2011) | **Editing**: the process of making text "complete, accurate, correct, comprehensible, usable, and appropriate for the readers" (p. 8). |
| | **Technical editing**: involves subject matter that requires specialized expertise and/or subject matter that requires the editor to "analyze, explain, interpret, inform, or instruct" (p. 11). |
| Amare, Nowlin, & Weber (2011) | **Technical editing:** the process of communicating complex information to audiences in understandable terms. |

According to these definitions, technical editing can be *technical* in multiple senses of the word.

## Rhetoric-Based Definitions

Technical editing tends to be defined rhetorically; that is, most technical communicators have defined the term in relation to a specific audience, purpose, and context.

| Greenberg (2010) | **Editing**: "a decision-making process, usually within the framework of a professional practice, which aims to *select, shape* and *link* content . . . to help deliver the meaning and significance of the work to its audience" (p. 9). |
| Tarutz (1992) | **Technical editing**: includes "any specialized subject that addresses a specific audience, has its own jargon, and whose approach is objective" (p. 4). |

Both definitions address the rhetorical situation. Much like Hayhoe (2010), Tarutz defines technical editing broadly to cover all senses of the term technical. In contrast, Dragga and Gong (1989) liken editors to artists who use rhetoric as an artistic tool:

[Editors] may take other writers' texts and help shape them for presentation to readers, or they may compose texts themselves, using their expertise in the craft of writing; they may weave together a number of works into a tapestry—an edition or collection of texts; they may design the actual aesthetic presentation of texts—typeface, type size, paper, etc.

(p. 11)

In this description of editor's work, writing and editing activities overlap in certain circumstances. Similarly, Norton (2009) acknowledges crossovers between writing and editing activities in developmental editing.

## Actor- and Activity-Based Definitions

Technical editing has been defined in terms of the actors and the activities performed.

---

Murphy (2010)   **Technical editing**: "the planning, analysis, restructuring, and language changes made to other people's technological or scientific documents in order to make them more useful and accurate for their intended audiences" (p. 1).

---

Some scholars (e.g., Haugen, 1990; Murphy, 2010) emphasize that an author revises his or her own document, whereas an editor edits other people's documents. Although editing differs from revising, the differentiation of actors' roles and activities may get blurred in practice, particularly in workplaces that rely on peer review and/or subject-matter expert (SME) review or in circumstances where the editor becomes a ghostwriter due to tight deadlines, tight budgets, or other situations.

## Discipline-Based Definitions

Editing means different things to different people. Weber (2010) sees significant overlaps between editing categories. Some fields, including technical fields such as medicine, limit the definition of technical editing to lower-level concerns such as fixing grammar and mechanics (i.e., copyediting)—tasks others attribute to literary editors (e.g., Boomhower, 1978, as cited in Haugen, 1990). Boomhower (1978) views technical editing in terms of clarity and concision; however, those technical communication constructs align closely with Einsohn's (2011) communication-driven copyediting criteria: "clarity, coherency, consistency, and correctness" (p. 3). Ironically, the concept of editing is further muddled by fields such as composition that interchange the terms "responding," "reviewing," and "commenting" with "editing" (Charlton, 2013, p. 103). Students may mistakenly conclude that proofreading = copyediting = editing.

## Levels-Based Definitions

In light of this confusion, it's not surprising the technical communication field appropriated Van Buren and Buehler's *Levels of Edit* (1980) to help define what editing entails. Although developed to streamline cost estimates at the Jet Propulsion Laboratory, the levels of edit essentially outline a menu of editing services. The five levels encompass nine types of edit: coordination, policy, integrity, screening, copy clarification, format, mechanical style, language, and substantive; the first level includes all types of edit, and level five includes only two types of edit: coordination and policy (Buehler, 1981).

The levels have been expanded to include usability (Soderston, 1985) and websites (Anderson, Campbell, Hindle, Price, & Scasny, 1998); condensed to two to four levels (Amare, Nowlin, & Weber, 2011); and refined to prioritize content quality, content worth, and cost-effectiveness (Nadziejka, 1995). Some have criticized the levels of edit for their prescriptive nature (e.g., Nadziejka, 1995) and omissions of various time-intensive tasks, such as editing second-language documents (e.g., Tarutz, 1992). Regardless of the number of levels of edit, many editors adopt a triage approach, devoting limited resources to the most critical areas first (e.g., Einsohn, 2011; Tarutz, 1992). In the classroom, the levels of edit can provide a framework to scaffold students' mastery of tasks or to structure students' workflow.

## Literature Review

In this section, I review the technical editing literature published since 2000 and connect that research to editing pedagogy. I begin with an overview of the technical editing profession and then shift to editorial roles, editing elements, and editing technology. I end the review with a discussion of editing pedagogy and curricula.

The technical editing profession originated centuries ago. Malone (2006) traces the emergence of "learned correctors"—language and/or subject-matter experts who helped publishers and printers correct texts—in response to technology-related errors and communication problems that occurred with the introduction of moveable type (p. 390). Malone describes the history of publication production processes—including the use of markup systems, the types of printing errors, and the errata correction methods—and the evolution of the correctors' work as technical editors. In a similar vein, Hayhoe (2010) outlines the history of technical journals, describes typical operating procedures in academic publishing, and discusses metrics such as acceptance rates and citation indexes. Warren (2010) picks up the history of technical editing from WWII and recounts how the editing literature has shifted from personal accounts to empirical studies.

The field has progressed considerably; however, more research on technical editing is needed (Eaton, 2010). In the field's major peer-reviewed journals, I found 16 empirical studies on editing published between 2000 and 2017 (e.g., Norman & Frederick, 2000; Kreth & Bowen, 2017); in that sample period, *Intercom* published at least 35 editing-related pieces, of which at least 11 appeared in columns dedicated to editing (see Figure 2.1). The peer-reviewed studies involved sample sizes from three to 580 and methods ranging from observations to interviews. Textual analyses (e.g., discourse analysis, content analysis) were the most common method, used in seven studies, followed by surveys, which were used in six studies; some studies used multiple methods. Predictably, the smallest sample sizes were associated with interviews ($N = 3$ to 20) and the largest sample sizes, surveys ($N = 176$ to 580); textual analyses ($N = 11$ to 105) fell between the extremes.

As of the third quarter of 2018, no additional technical editing articles had been published in the major technical communication journals. This volume adds four empirical studies to the discipline's body of empirical knowledge, an increase of 25% since 2000, as well as three theoretical pieces. The following sections summarize the discipline's current understanding of technical editing.

## Editing Profession

Kreth and Bowen (2017) surveyed 253 self-identified technical editors to determine what their jobs entailed, how their work is perceived, and what their career paths look like. The authors did not calculate interrater reliability; the study results are descriptive and the qualitative details presented were selected for context, reader interest, and usefulness. The survey reports on participants' editing-related activities (e.g., responding and proofreading), editing focus (e.g., accessibility and usability), editing tools (e.g., FrameMaker and LaTeX), collaborative editing experiences (e.g., working with second-language authors), career preparation and professional development (e.g., education and first jobs), editor personality traits (e.g., flexible and detail-oriented), and editors' perceptions of the value of technical editing work (e.g., reputation and salary). As did Dayton (2003), Kreth and Bowen (2017) found people working in technical editing jobs in diverse fields, often under job titles without the terms *technical* or *editor* in the title. Regardless of the job title, technical editing is rarely entry-level work (Corbin, 2010).

## Workplace Editing

Building on Thompson and Rothschild's (1995) workplace study of three editors, Bisaillon (2007) conducted a case study of six professional editors to determine what revision strategies they used in the workplace. She found the editors typically read the documents from one to three times, which suggests that professional editing is a linear process; the editors used several problem-solving strategies: rereading, reflection, immediate search, postponement of solution, and tentative solution.

Relying on their knowledge, the more experienced editors immediately corrected problems nearly 75% of the time; when solutions weren't apparent, the experienced editors tended to reflect on the problems (Bisaillon, 2007). In contrast, the less experienced editors tended to search for solutions, but their searches only proved helpful about 53% of the time (Bisaillon, 2007). Bisaillon concluded that reflection and rereading are the most effective problem-solving solutions. These findings may help editing students develop effective editing strategies.

A study of authors' perceptions of editing in the workplace contradicts previous reports of adversarial author–editor relationships; Eaton, Brewer, Portewig, Davidson, and Portewig (2008b) surveyed 449 authors and found 76% reported working relationships that ranged from good to excellent, and only 3% reported negative relationships. Most participants perceived editing to be "crucial to maintain credibility, accuracy, and quality" (Eaton et al., 2008b, p. 129). In the classroom, editing students would benefit from discussions on how interpersonal skills impact author–editor relationships.

## Editing Tests

The Society for Technical Communication (STC) offers certifications for professional technical communicators; the reviewing and editing portion of the foundation-level exam assesses knowledge of editing processes, style guidelines, levels of editing, usability, and of course grammar and mechanics (Baehr, 2016). Although certification provides evidence of competency, technical editors may face another employment hurdle: an editing test. Research on editing tests (Boettger, 2012, 2014) provides insight on employers' expectations and skills that should be incorporated into editing curricula.

In the first of three studies on editing tests, Boettger (2014) determined the characteristics of editing tests (e.g., formats, administration and evaluation practices). In the second study, he analyzed the content of 55 editing tests from nine industries that use the assessment in hiring technical editors or technical writers; this study looked at the types and distribution of errors. Of the 71 types of errors identified, most were categorized as grammar and mechanics and style. On average, each test featured 23 types of errors with about 66 total errors. The three most predominant errors were wrong word, misspelling, and unnecessary or missing capitalization, in that order; the errors were weighted by frequency and dispersion.

In the third study, Boettger (2014) surveyed technical communicators to determine their perceptions of 24 errors. He found participants were most bothered by content errors; however, in the sample of editing tests, content errors occurred less frequently than errors in grammar and mechanics, style, and punctuation. Participants were least bothered by style errors, yet style errors were the second most frequently occurring error in editing tests. Boettger attributes the differences partly to context; the editing tests were designed to assess one's knowledge of style guides.

When comparing the types of errors in medical editing tests with those from other fields, Boettger (2012) found the error distribution patterns differed considerably, particularly in respect to style and passive voice. He identified 60 types of errors in the medical editing tests, with errors in style, grammar and mechanics, and content appearing most frequently, in that order. Boettger speculated that style errors might be "easier to assess" or conversely "the most subjective error type and therefore the most difficult to assess" (p. 102). Aside from Boettger (2012, 2014), the literature provides limited advice on taking editing tests (e.g., Hart, 2003) or on assessing technical editing skills in the classroom. We talk about what to teach and question whether grammar exercises work (MacNealy, 1999), but without empirical evidence to support editing assessment practices, it's hard to ensure that students learn the concepts well enough to transfer the skills to other editorial situations.

## Checklists

When customized for each project or client, editing checklists may help improve factors ranging from an editor's efficiency to the document's structure and consistency (Schrank, 2013). Well-constructed checklists illustrate the attention to detail required for effective editing and reinforce the reason for performing multiple editing passes. Although Schrank (2013) says editing checklists are intended to help "an experienced editor keep track of details for particular types of documents" rather than "to teach someone how to edit a document," others have found checklists to be useful classroom tools (p. 166). Editing checklists can help students improve their editing skills if students are taught to "internalize what would appear on a checklist" and to integrate the checklist in a multipass-editing process rather than use the checklist as a one-pass task completion record (Albers & Marsella, 2011, p. 64). Albers and Marsella (2011) posit that multipass edits should reduce cognitive loads and improve editing consistency. Improvements in consistency should lead to improvements in quality.

## Editorial Ethics

Amare and Manning (2009) interviewed three academic journal editors from the fields of technical communication and business communication to understand how editors handle various ethical dilemmas. Their study builds on Allen and Voss's (1998) study of editorial ethics in the workplace. Amare and Manning (2009) posed six questions designed to interrogate values identified by Allen and Voss, such as honesty, quality, and legality. Amare and Manning use Peirce's epistemological ethic as a theoretical framework for analyzing the interview responses and they present their results in the form of anonymized scenarios. The scenario topics range from plagiarism and peer review to editorial decisions, guest editors, and author status—all of which could serve as the basis for classroom discussions on editorial ethics. Amare and Manning's results seem to contradict an earlier study by Dragga (1996) that suggested technical communicators tend to work alone without formal ethical guidelines.

### Editorial Roles

Like many professionals, editors function in multiple capacities and may perform clerical to managerial tasks (Cundiff, 2017; Kreth & Bowen, 2017). Editors' perceived status and job titles vary by industry and workplace situation.

### Status

The value of editors and editing has persisted as a theme in the literature for decades. Lanier (2009) stratifies editors as high-status (e.g., editors in environments where publishing is the primary activity) and low-status (editors in environments where publishing is a secondary activity). In workplace contexts, editing has been described as "a subsidiary, if not marginalized, activity" (Dayton, 2004a, p. 218) and a "lower status . . . collateral function" (Buttram, 2016, p. 10). According to Dayton (2003), editing in the 1990s was "a de-specialized, distributed function in most technical communication workplaces" (p. 192). Others have described editing as "a high-level, strategic process that requires a special kind of expertise" (Haugen, 1991, p. 62).

The tendency to undervalue editorial work also exists in academic contexts. For instance, tenure and promotion guidelines usually elevate authors over editors (Amare, 2009). Amare contends this value hierarchy "stem[s] not only from our privileging invention but from our belief that authors own their texts" (p. 195).

Writing and editing are "distinct and separate tasks" that require different communication skills (Corbin, Moell, & Boyd, 2002, p. 295). Positioned in the middle of the communication situation, editors function as mediators between authors and readers (Buehler, 2003). When performed well, editors' work is invisible to readers. As a result, it is hard to quantify the value of editors' work, which means technical editors must promote themselves as professionals who add value and contribute to quality control and business goals (Greenberg, 2010; Kreth & Bowen, 2017). Greenberg (2010) maintains "we need to develop new ways of evaluating the added value of 'invisible' intermediary work [i.e., editing]" (p. 18).

Several scholars believe an editor's ability to adapt to technological changes and to provide "skilled human intervention" adds value (Greenberg, 2010, p. 13) and perhaps even elevates the editor's status. For example, in a retrospective case study of a government laboratory, Lanier (2009) reflects on how technology changes can impact editorial authority; when required to use unfamiliar electronic editing tools, the authors relied on the editors' technological expertise. Editing technologies—and editors' roles—evolved further with the shift to modular content development. Amare argues that the use of content management systems "is elevating the technical editor's role in document production" (2009, p. 182), and Corbin, Moell, and Boyd (2002) believe technical editors' specialized skills "will become more critical as information is combined and recombined from a variety of sources into a variety of information deliverables" (p. 296). Similarly, Kreth and Bowen (2017) contend that "rather than shrinking in importance and relevance, the work of technical editors continues to find new and diverse homes where technical editors' skills and strengths pay dividends" (p. 254).

In workplace and classroom contexts, we need to dispel characterizations of editors as "grammar janitors" (Rude, 2006, as quoted by Amare, 2009). We must emphasize the value of editorial work in both human and financial terms, yet also acknowledge how editorial technologies impact power structures and editorial authority (Lanier, 2009). To gather empirical evidence that quantifies the value of editorial work, we might focus on how copyeditors to editors-in-chief improve user experience and quality control.

## Quality Control

Editors play a key role in maintaining quality control. Some organizations recognize that quality information improves user experience. For that reason, IBM implemented an Editing for Quality (EFQ) process in which nine quality characteristics grouped in three categories (see Table 2.1) are rated on a scale of 100 by two editors; their evaluation process requires an agreement score of at least 85 (Wilde, Corbin, Jenkins, & Rouiller, 2006). The process "not only improves the quality and consistency of information, but also increases the skills of both writers and editors" (Wilde et al., 2006, p. 439).

Other organizations need evidence of editors' contributions to quality and the bottom line. To make a case for editors' value, Corbin, Moell, and Boyd (2002) mapped technical editing tasks (i.e., comprehensive editing, usability editing, and copyediting) to those of software testers (i.e., system testing, usability testing, function testing) and argued that "by providing quality assurance through content editing, technical editors add value to the information development process and help to give users the quality content that they deserve" (p. 287). They contend that neither self-editing nor peer review provide the same level of quality as technical editing. Greenberg (2010) concurs and warns

TABLE 2.1 IBM's information quality characteristics (Wilde et al., 2006, p. 440)

| Category | Easy to use | Easy to understand | Easy to find |
|---|---|---|---|
| Quality characteristic | task orientation accuracy completeness | clarity concreteness style | organization retrievability visual effectiveness |

"the absence of editing has important consequences for communication and the creation of meaning" (p. 10).

Within and beyond the classroom, a standardized editing process, such as the EFQ, can be useful for establishing information quality benchmarks. Metrics and additional empirical evidence can strengthen the case for editors' value and their contribution to the bottom line. Current literature provides little guidance on how to bring benchmarks into the classroom.

### Editing Elements

Editorial work not only requires language expertise but also theoretical grounding, audience awareness, and constructive feedback.

### Editing Theories

Rhetoric underpins recent editing theories. Dragga and Gong (1989) view rhetorical theory as "the basis or philosophical foundation for the editor's judgements" (p. 11). Likeminded, Buehler (2003) presents a situational editing theory that requires editors to be flexible and aware of the big picture. Buehler's (2003) theory of editing complements traditional programmatic or rule-based approaches to technical editing; her theory requires applying grammatical rules and house style within the context of the rhetorical situation. According to Buehler, "no amount of correct grammar and concise writing will ensure that the information is transmitted effectively unless the needs of the audience are correctly assessed and fulfilled" (2003, p. 461). To that end, Flanagan (2015) proposes a user experience approach to editing modular content for dynamic reuse. Others have noted the difficulties of editing modular content (e.g., Albers, 2000; Swarts, 2010; Williams, 2003).

Hands-on methods work well for introducing students to editing theories. For instance, Masse (1985) encourages students to develop editing theories based on research and their experience using Van Buren and Buehler's (1980) levels of edit. Buehler's (2003) theory is suitable for editing students who have developed editorial judgment and confidence in their editorial decisions. Editing students of all experience levels could benefit from Popham's feminist approach to editing (see Chapter 5). Popham's theory calls for "respect[ing] the work and words of others, not just as a matter of correctness, but as a situation of cooperation, respect, and fairness" (Chapter 5, p. 107).

### Audiences

Technical editing textbooks emphasize the importance of audience awareness and provide guidelines for editing content for global audiences with various cultural expectations. Published before the World Wide Web existed, Tarutz's (1992) textbook allots only four pages to editing for international audiences compared with more recent textbooks that devote entire chapters to the topic. Amare et al. (2011) outline the G.I.L.T. process (globalization, internationalization, localization, and translation), and demonstrate how common words vary across five versions of English. Maylath (2011) includes a localization checklist and a list of cultural editing concerns.

Empirical studies have shown that culture affects authors' receptiveness to and interpretation of editorial comments (Eaton et al., 2008a; Mackiewicz & Riley, 2003). In this volume, St.Amant (Chapter 8) discusses how content may be edited for direct use (end users) or indirect use (translators and localizers) and recommends the preemptive strategy of "editing in a way that addresses both direct and indirect audiences" in case businesses expand to global markets. Editing students must be encouraged to extend their view of audience to include global audiences as well as audiences with disabilities who may access texts through screen readers and other technologies. In addition, students must consider the nonhuman audiences run by artificial intelligence, which support dynamic information architectures. To effectively communicate with bots, students must be familiar with metadata, extensible markup language (XML), and search engine optimization (SEO).

## Editorial Comments

Reviewers and editors routinely provide feedback to authors; the feedback may be communicated verbally or through handwritten or electronic comments (e.g., copyediting marks, text scrawled in the margins or on sticky notes, electronic annotations). Tactfully communicating constructive criticism to authors in ways that motivate them to comply with editorial requests can be challenging (Buttram, 2016; Eaton et al., 2008a, 2008b; Mackiewicz & Riley, 2003).

Applying linguistic research on politeness to editor–writer communications, Mackiewicz and Riley (2003) outline eight strategies for balancing clarity and politeness in editorial comments to authors; the commenting strategies range from opinions to hints and span the direct–indirect continuum. The commenting strategies can be paired with six types of downgraders, such as hedges, that are intended to increase the perceived politeness of editorial comments. Mackiewicz and Riley recommend using downgraded, indirect comments; however, others disagree with that approach (e.g., Buttram, 2016; Eaton et al., 2008a).

Some scholars find direct comments with "payoff statements" or payoff statements followed by specific suggestions work best (e.g., Buttram, 2016; Eaton et al., 2008b). Only two of Mackiewicz and Riley's (2003) comment suggestions "performed as predicted" in Eaton et al.'s 2008 survey of more than 400 authors from various countries; the study also revealed that 95% of the authors contact editors if they do not understand a comment (2008b, p. 135). Editors should also consider authors' communication preferences. In Eaton et al.'s (2008b) study, the researchers found 23% of the participants prefer electronic comments and 23% prefer electronic comments followed by face-to-face meetings.

Commenting strategies should be selected with culture and context in mind; nonnative English speakers interpret comment phrasing differently than native English speakers (Eaton et al., 2008a, 2008b). Eaton et al.'s (2008a) survey data showed that "nonnative speakers were more than twice as likely to not typically follow copyediting comments (15% versus 6%) and proofreading comments (20% versus 9%) than native speakers of American English" (p. 144). Both groups were equally likely to comply with comprehensive editing comments, but younger participants were more likely than older participants to follow comments from managers; participants' receptiveness to certain comment types varied, as did their perceptions of editing experiences—older participants reported more positive experiences (Eaton et al., 2008a).

In circumstances where volunteer editors deal with user-generated content, the type of comment may have less impact on writers' motivation to revise. Mackiewicz (2011) studied a social media review site and discovered that reviewers were equally likely to revise in response to welcome messages as to messages with editing comments. The volunteer editors used various politeness strategies (e.g., compliments, hedges, and hints) to recommend substantive edits (e.g., defining terms and correcting omissions, inconsistencies, technical discrepancies) as well as other levels of edits: screening, policy, format, and language. Her 2014 follow-up study, which emphasized content quality rather than politeness strategies, showed screening comments were "strongly associated with updated reviews" but praise didn't seem to affect whether reviews were revised (p. 436). She attributes praise's weak "motivating effect" to the volunteer editors' lack of authority to "mandate changes" (Mackiewicz, 2014, p. 429). Similarly, Eaton et al. (2008a) found that, in the workplace, authors were more likely to make changes when time was not a constraint and when the editor was a manager.

Although indirect comments are considered more polite (Mackiewicz & Riley, 2003), students in one undergraduate editing class used a direct approach in 75% of their comments (Albers & Marsella, 2011). Most of the students' comments related to paragraph-level errors (61%), and the rest related to sentence-level errors (30%) and document-level errors (8%); only 9% of the comments explained the problems identified (Albers & Marsella, 2011). Albers and Marsella (2011) recommend teaching students to write comments with payoff statements that explain the rationale for suggested edits (p. 65). Another factor to consider when teaching students how to provide editorial feedback is authors' technology skills.

Boettger (Chapter 3) analyzed four psychological dimensions of graduate students' editorial comments. The corpus analysis showed the student editors used high percentages of clout (87.05%), emotional tone (77.53%), and analytical thinking (67.52%) in their comments but low percentages of authenticity (27.00%). The nature of technical writing may help explain the low authenticity dimension. Boettger found social variables, such as gender and native language, had little influence on how the student editors commented. During classroom discussions, students should consider how the sentiment of editorial comments impacts author–editor relationships.

Another important interaction is human–information interaction, which deals with "how people find, interpret, and use information to best achieve their goals" (Albers, 2008, n.p.). Students need to understand how audiences use texts in order to provide effective editorial comments. Yet, an analysis of 56 graduate student papers shows a tendency to perform line edits rather than global, document-level edits (Albers, Chapter 6). This finding is consistent with Cuppan and Bernhardt's (2012) study of peer review in the pharmaceutical industry. Neither the students nor the reviewers considered how the audience would use the information. Albers contends we need to teach students to edit for human–information interaction (or audience–information interaction) at the developmental editing stage.

### Editing Technology

In today's digitally connected world, people communicate daily using smart phones and mobile computer devices with Wi-Fi or high-speed internet connections. These technological developments have also changed editorial practices, including preferred editing modes, tools, and technologies.

### Editing Modes

Several studies document the transition to electronic editing, also known as e-editing (Lanier, 2004) and softcopy editing (Coggin & Porter, 1993). Dayton (2003) defines electronic editing as "any method of suggesting, showing, or making changes to an electronic copy of a text authored by someone else" (p. 195).

Electronic editing met with considerable resistance from editors and writers; early adopters tended to work from a distance (Dayton, 2003). Initially averse to change, many editors and writers balked at disruptions to well-established paper-based editing processes and clung to existing markup conventions. ASCII-based electronic markup systems that resembled paper-based markup conventions emerged; some of the electronic markup systems were designed to work across operating systems and for both print and electronic output modes (Kuhlenschmidt & Mosby, 2001). People used a variety of electronic markup methods ranging from highlighting and strikethrough to special characters and formatting to Microsoft Word's track changes and commenting features (Dayton, 2004b).

Resistance to electronic editing was attributed to factors such as computer-related ailments, technology limitations, blurred authorship lines, and requirements to preserve records of editorial changes (Dayton, 2003; Kuhlenschmidt & Mosby, 2001; Lanier, 2004). Over time, improvements to computer hardware (e.g., higher resolution monitors and ergonomic keyboards) reduced eye strain and repetitive stress injuries, and advancements in word processor capabilities resolved concerns over preserving the author's original text (e.g., Microsoft Word's track changes and commenting features). The diffusion of electronic editing was further hampered by limited access to specific operating systems (e.g., UNIX) and expensive software such as FrameMaker (Dayton, 2003, 2004b).

Among the biggest obstacles were editors' and writers' attitudes toward technology and their perceptions of editing practices. In a three-part series on electronic editing, Dayton (2003, 2004a, 2004b) reports the results of a 1999 survey of STC writer-editor members' editing practices and attitudes and recounts the findings from interviews with 20 technical communicators at five companies. His goal was to identify the differences between companies using hardcopy editing versus electronic editing.

At the time, most of Dayton's survey participants used both paper and electronic editing; hardcopy editing was used by more than half of the participants (54%); of those editing electronically, about two-thirds automated editing tasks through the use of macros (Dayton, 2003). Participants' views of electronic and hardcopy editing varied from person to person and from workplace to workplace (Dayton, 2004b). Twelve years later, a 2011 survey of technical editors conducted by Kreth and Bowen (2017) found that 75% of the participants perform copyediting and proofreading tasks entirely electronically.

Responding to Dayton's (1998) call for research on editors' and authors' attitudes toward electronic editing, Lanier (2004) interviewed five authors in a government research laboratory to gauge their perceptions of electronically edited documents. Some authors who were resistant to electronic editing needed detailed written instructions on how to use Microsoft Word's reviewing features (Lanier, 2004). Similarly, Dayton found that some editors weren't aware of Word's track changes and commenting features whereas others chose not to use them because of usability and compatibility issues (Dayton, 2004a, p. 215).

For Lanier's (2004) study, editors' use of Microsoft Word's track changes and commenting features increased the frequency of collaboration, eased authors' fears of editors changing texts' meanings, and reduced the time spent on line-by-line reviews; even so, some authors preferred to print the edited documents with tracked changes showing. Lanier (2004) noted the "authors seem to perceive the electronic comments differently and more beneficially than they had the hand-written notes" (p. 531). He attributed this finding to factors such as ease of reading and clear connections between the comments and the corresponding text. In contrast, Eaton et al. (2008a) found "those born in the U.S. more frequently preferred to receive comments on paper" (p. 145).

In certain circumstances, the affordances of paper are superior to electronic editing technologies. For example, many people prefer hardcopy editing for long documents because the structure is easier to see in hardcopy and the cognitive load is reduced (Dayton, 2003; Kuhlenschmidt & Mosby, 2001). Some people prefer hardcopy editing for documents heavy in numerical data, equations, and technical graphics (Dayton, 2003). Aside from the nature of documents, reasons for favoring paper-based editing range from portability and ergonomics to ease of spotting errors and quality control (Dayton, 2003, 2004a). Although touch screens and stylus tools mimic the affordances of paper and support hardcopy-style markup, when Dayton conducted his survey those technologies had not been widely adopted for editing purposes (Dayton, 2003, 2004a).

Invoking Roger's (1962, 1971) diffusion of innovations theory and the concept of reinvention, Dayton (2004a) theorizes that

> the diffusion of electronic editing in technical communication has advanced mainly by reinvention, gradually, erratically, as authors and editors adapt their software tools to the tasks of marking edits and inserting queries in electronic texts—or avoid the perceived problems with on-screen markup by using paper for the task.
>
> (p. 215)

Other factors influencing editing methods include company policies, workplace cultures, and access to compatible tools and technology; "different workplace contexts ... dictate different editing choices" (Dayton, 2004a, p. 217). More than a decade later, a virtual focus group of 27 technical editors from six countries concurs that editing tools "vary according to industry demands and requirements, and author and organizational preferences" (Lanier, Chapter 7, p. 142).

Whether in the workplace or in the classroom, people (i.e., editors, authors, reviewers, and students) need to be trained to use electronic editing tools. Although advanced word-processing features such as track changes and insert comment have been available for years, many people are unaware of or unfamiliar with these and other editing technologies (Lanier, 2004); even students who are digital natives—those who have grown up with computer technologies—"are often surprisingly unaware of available tools in word processing" (Rude, 2010, p. 62). Ongoing training in the use of editing tools is needed as technology evolves. As Lanier and Melonçon point out in this volume, students should expect to learn new technologies and apply those foundational skills to different workplace situations (see Chapters 7 and 9).

## Editing Practices

Editing practices are functions of workplace contexts, management philosophies, information infrastructures, and available tools and technologies. In particular, single-sourcing practices (creating modular content once for multiple reconfigurations and reuses) and content management technologies (database systems for coordinating the storage and publication of modular information) have impacted editorial roles and workflows. Depending on the information architecture of a content management system (CMS), the editor's role may or may not be prioritized. In some workplaces, the roles of writer, editor, and subject-matter expert are blurred by technology (Kramer, 2003). When layout and design tasks are automated and documents are compiled on demand from database content, writers may move into information management-technologist roles, and editors—if they haven't been replaced by writers—may lose control over the final document (Albers, 2000; Kramer, 2003).

Though some have connected technology to the disappearance (Greenberg, 2010) or replacement (Kramer, 2003) of editors, others see technology as providing new opportunities for editors (Kreth & Bowen, 2017)—complex opportunities that require technical editors' expertise throughout the content production process (Albers, 2000; Amare, 2009; Flanagan, 2015). Dynamic documents require self-contained information chunks and an underlying structure that provides "a coherent communication path to the reader" (Albers, 2000, p. 194). A CMS cannot effectively facilitate communication unless a human editor "control[s] the formation of data modules and create[s] the appropriate metadata, tags, or semantic labels" (Amare, 2009, p. 190). Without a technical editor, the information may not be usable (Albers, 2000)—garbage in; garbage out.

Just as editors must use their skills to "align the document's message with user needs" (Amare, 2009, p. 185), students must be taught to develop and edit content to meet users' information needs in Web 2.0 (user-generated content), Web 3.0 (semantic-based automated content), and subsequent Web X.0 environments where content is interactive and dynamic (Albers, 2000; Greenberg, 2010; Mackiewicz, 2014). Using a humanistic approach, Bridgeford (Chapter 4) teaches students to envision themselves as editors working with CMS technologies.

## Editing Pedagogy and Curricula

Prior to 1975, technical communication instructors had limited resources for teaching technical editing; they relied on general editing books and journalism textbooks until several technical editing textbooks were published (Warren, 2010). Currently available top-selling technical editing textbooks (e.g., Amare et al., 2011; Murphy, 2010; Rude & Eaton, 2011; Tarutz, 1992) are well suited for print-based publishing but provide limited coverage of multimodal-based publishing issues (Lang & Palmer, 2017). Although instructors now have a variety of resources, few articles have been published since 2000 that primarily address editing pedagogy issues (e.g., Charlton, 2013; Gaitens, 2000; Lang & Palmer, 2017; Norman & Frederick, 2000) or modern publishing technologies (e.g., Kreth & Bowen, 2017).

Universities have emphasized traditional editing skills (e.g., copyediting, grammar, markup), and some scholars (e.g., Lang & Palmer, 2017) have called for updated technical editing curricula that reflect modern workplace practices and align with employers' needs. Lang and Palmer (2017) analyzed job ads, technical editing literature and textbooks, and university course descriptions and concluded that today's technical editors need multimodal content skills. In light of Kreth & Bowen's (2017) findings that few editors edit audio or video, Melonçon (Chapter 9) questions the need for those particular digital editing skills. Based on her analysis of programmatic materials from 40 institutions, Melonçon suggests that editing curricula should instead place more emphasis on editorial ethics, visual design, and global issues.

To better prepare for workplace and freelance career paths, editing students need knowledge in areas beyond communication, such as technology, business, and user experience (Kramer, 2003; Kreth & Bowen, 2017). Editing principles are well established and likely to remain stable, but editorial skill sets must evolve with ever-changing technologies (Tarutz, 1992).

Students can benefit from opportunities to apply their developing knowledge and skills to real communication problems; however, students may struggle in various ways. For example, NC State University students who participated in writing and editing internships in the English department experienced socialization problems; over five semesters, Gaitens (2000) identified problems related to independence, isolation, failure to ask questions, difficulties with adapting to different ways of writing and communicating, and a lack of awareness of workplace cultures.

Similarly, Norman and Frederick (2000) recount problems integrating technical editing teams into an engineering design project; in a three-year experiment, editing students served as project editors for the engineering students who worked on industry-based integrated product teams (IPTs). Norman and Frederick configured the editing teams in three ways: editors as full IPT members, editors as one independent team, and three editing teams with liaisons to the IPTs. Each of the three approaches had different advantages and disadvantages, many of which were functions of the editing students' class size, skill levels, and ability to work well in teams. The major challenges included managing teamwork issues and balancing the learning curve, the workload, and the deadlines. Norman and Frederick concluded "the ideal situation for fully integrating editors into IPTs would be to have an official 'IPT Class' that technical communication graduate students and very advanced undergraduates could take as a form of internship or independent study" (p. 184). However, Gaitens' (2000) results suggest socialization problems would arise.

Charlton (2013) addresses problems that stem from preconceptions of editing and editors' roles. He outlines the theoretical rationale for a dual-level technical editing course he based on Amare, Nowlin, and Weber's (2011) textbook.

Charlton (2013) provides a syllabus and critically reflects on the choices made within the context of institutional constraints and programmatic needs.

Lang and Palmer (2017) report similar challenges in redesigning their editing course to better reflect editing in today's workplace. Informed by their research, they incorporated modules on editing audio, video, and websites. In a subsequent semester, they refined the course based on students' feedback; additional readings on grammar and mechanics were added and changes were made to accommodate students' inexperience with various technologies. Although the revised technical editing course aligned better with the modern workplace, Lang and Palmer conceded that the course had evolved into a survey course. They suggest integrating editing tasks into existing technical communication courses, adding separate courses in grammar and multimodal editing to the curriculum, and creating a sequence of editing courses.

## Conclusion

Based on decades of technical communication lore, the technical editing literature covers a range of topics. Although the literature is evolving into a body of tested knowledge, our technical editing knowledge is limited in scope. The majority of the empirical research has been conducted on editorial comments, electronic editing, editing tests, and quality control. These studies represent the work of a handful of technical communication scholars. We must engage more scholars in more research that will inform editing pedagogies for the modern classroom. The literature points to a future in which the role of artificial intelligence and computer technologies will become increasingly integral to editorial work. Much remains to be investigated, particularly the emerging and ever-changing publishing technologies.

### Open Questions

Future empirical inquiries might begin with one or more of the following open questions.

### Workplace Editing

- How do people edit? "Which operations are automatic from the start? Which ones become automatic with practice?" (Bisaillon, 2007, p. 319).
- "Which [editing operations] will always require reflection and judgment?" (Bisaillon, 2007, p. 319).
- What do editing practices look like in different contexts (e.g., industry, nonprofits, education, government, freelance; Bisaillon, 2007)?
- What skills do editors need for editing content that is single-sourced, dynamically generated, or user generated (Kramer, 2003)?

- Where does technical editing fit best within iterative project management workflows such as Agile?
- How cost effective is editing at late stages of the writing process? How does writing quality compare when editing occurs early in the writing process versus late in the writing process (Eaton et al., 2008b, p. 135)?
- What editing concerns do volunteer editors prioritize compared with professional editors (Mackiewicz, 2011)?

## Editing Tests

- What types of errors are specific to technical communication or occur more frequently in technical communication (Boettger, 2014)?
- What types of errors are most bothersome to technical communicators (Boettger, 2014)?
- Are editing tests valid assessment instruments for hiring purposes? For classroom purposes?
- What alternatives to editing tests might work better (Melançon, Chapter 9)?
- How do classroom assessments align with workplace assessments?
- How might benchmarks be used in the classroom? What benchmarks should be used?

## Audiences

- What are the best practices for editing and responding to documents written by nonnative English speakers?
- How do cultural, language, and demographic factors impact author–editor relationships and authors' perceptions of and responses to editing (Eaton et al., 2008a)?
- What "models, processes, or conceptions of editing" exist among different cultural and demographic groups? How do editorial practices differ between groups (Eaton et al., 2008a, p. 158)?
- How do personas affect students' understanding of audience?

## Editorial Comments

- "What's the best way to phrase editorial comments?" (Boettger, 2014, p. 229).
- Why do "women and men [interpret] editorial comments differently" (Eaton et al., 2008a, p. 158)?
- Why are some groups "significantly more likely to follow proofreading and copyediting comments than their counterparts, but there is no significant difference in how often these groups follow comprehensive comments?" (Eaton et al., 2008a, p. 158).

- How do "authors actually address comments (as opposed to . . . how they believe they address comments)?" (Eaton et al., 2008b, p. 136).
- How do students' commenting abilities change as their editing skills develop (Albers & Marsella, 2011)?
- How do we teach students to analyze texts at the document level in order to identify and comment on high-level communication problems (Albers, Chapter 6; Albers & Marsella, 2011)?

## Technology

- What does technical editing look like in a Web 2.0, Web 3.0, and Web X.0 world? How, if at all, do editing strategies differ (Greenberg, 2010)?
- What are the best practices for electronic editing? Do these practices work for all populations and workplaces (Lanier, 2004)?
- What are the best practices for cloud-based editing?
- Which editing tasks can be automated? Which editing tasks require humans?
- How can editing technologies be improved?

## Editing Pedagogy and Curricula

- Which definition(s) of editing should we use for developing editing curricula (Melonçon, Chapter 9)?
- Which editing skills do students develop first? Which skills lag? What is the sequence of skills improvement? (Albers & Marsella, 2011).
- How do we "teach multipass editing in a manner that motivates students to use it" (Albers & Marsella, 2011, pp. 64–65)?
- Should we still teach hardcopy editing (Lang & Palmer, 2017)?
- How do we teach audio, video, and accessibility requirements (Lang & Palmer, 2017)?
- How can editing students be prepared for internship socialization issues (Gaitens, 2000)?
- How can error patterns (e.g., types, dispersion, and frequency) inform technical editing pedagogy and editing practices (Boettger, 2012)?
- Which editing strategies (e.g., make immediate corrections, reread, reflect, research solutions) work best for students (Bisaillon, 2007)?
- How should we "prepare students for careers as freelance and nonfreelance technical editors?" (Kreth & Bowen, 2017, p. 254).
- What is the most effective programmatic approach to teaching the myriad complex skills (e.g., technology, grammar, editing judgment, interpersonal communication) required of editors? Prerequisites, a series of courses, internships, independent study, etc.? (Kreth & Bowen, 2017; Norman & Frederick, 2000).

- How might editing and editing technology skills be scaffolded throughout technical communication programs or within editing courses? (Kreth & Bowen, 2017; Melonçon, Chapter 9; Norman & Frederick, 2000).
- Which courses should be prerequisites or corequisites (e.g., project management, computer courses, grammar, intro to editing, user experience, content strategy, etc.)? Which skills should be prerequisites? (Kreth & Bowen, 2017; Norman & Frederick, 2000).
- How much class time should be devoted to comprehensive editing, client-based projects, project management workflows (e.g., Agile), collaborative editing, structured languages (e.g., DITA), technology literacy, etc. (Melonçon, Chapter 9)?

## Pedagogical Applications

Although many questions remain open, we do know that technical editors will encounter many technology learning curves throughout their careers. The internet has changed how editing activities are performed, as well as editors' and authors' relationships with audiences, audiences' expectations, and audiences' behaviors. Increasingly, employers are expecting technical communicators to be "a magical purple squirrel employee that fits all needs and knows all software" (Lanier, Chapter 7, p. 140).

While it's not feasible to train students to know and do everything, we can prepare them to meet many industry needs. Technical editing students should be taught to approach editing work as a complex communication problem that requires strategic assessment; ethical, audience-centered solutions; and targeted attention to detail. In other words, students should analyze the writing situation and triage the text before fixating on grammar clean-up. Educators should stretch students' perceptions of technical editing and help students embrace a problem-solving mindset. In addition, educators should socialize students for collaborative work that demands strong interpersonal skills, technical aptitude, and flexibility. Students should be exposed to—or at least aware of—current editing tools and technologies, information architectures, and project management styles.

As editors, students must consider how people interact with and use information, and students must anticipate and advocate for the needs of culturally diverse audiences—direct and indirect. Beyond the classroom, editors must balance those audience needs with workplace constraints (e.g., time, money, labor, technology). They must also weigh the ethical implications of their editorial choices and be aware of potential legal ramifications—a seemingly harmless mistake, such as a misplaced or missing comma, can result in a rare but costly lawsuit.

In the classroom, personas may be a useful tool for helping students to envision audiences—from bots to nonnative English speakers—and their information needs. To help editing students understand structured writing and controlled language, technical communication educators could collaborate with foreign language educators to create an assignment in which technical editing students edit a text for translation and the foreign language students translate the text into another language. Afterwards, both groups of students could discuss their experiences and reflect on any communication pain points. Similar collaborations could be arranged with law students to convey the perils of miscommunication.

Equally important are collaborations with industry. Educators should continue to communicate with employers in the technical communication industry to monitor trends and assess training needs. Though employers may seek elusive "magical purple squirrels" for employees, they will find that well-educated technical editors can work magic on communication problems.

## Pedagogical Practicalities

- Students must understand the distinction between editing and revising: technical editors edit other people's work; authors revise their own work.
- Student editors must broaden their understanding of audience to include people with disabilities, people from diverse cultural backgrounds, people who speak languages other than English, and nonhuman bots run by artificial intelligence.
- Educators should introduce a triage approach to editing, not only to help students identify high-level communication problems but also to help them develop skills in time management and resource allocation.
- Educators should emphasize editing for audience–information interaction.
- Editors will be on the cusp of technology learning curves throughout their careers and must be able to transfer and adapt their existing knowledge to new situations.
- Editors (re)solve communication problems.

## References

Albers, M. J. (2000). The technical editor and document databases: What the future may hold. *Technical Communication Quarterly, 9*(2), 191–206.

Albers, M. J. (2008). Human-information interaction. *28th Annual International Conference on Computer Documentation*. Lisbon, Portugal, Sept 22–24, 2008.

Albers, M. J., & Marsella, J. F. (2011). An analysis of student comments in comprehensive editing. *Technical Communication, 58*(1), 52–67.

Allen, L., & Voss, D. (1998). Ethics for editors: An analytical decision-making process. *IEEE Transactions on Professional Communication, 41*(1), 58–65.

Amare, N. (2009). The technical editor as new media author: How CMSs affect editorial authority. In G. Pullman & B. Gu (Eds), *Content management: Bridging the gap between theory and practice* (pp. 181–199). Amityville, NY: Baywood.

Amare, N., & Manning, A. (2009). Examining editor-author ethics: Real-world scenarios from interviews with three journal editors. *Journal of Technical Writing and Communication, 39*(3), 285–303. doi:10.2190/TW.39.3.e

Amare, N., Nowlin, B., & Weber, J. H. (2011). *Technical editing in the 21st century.* Upper Saddle River, NJ: Prentice Hall.

Anderson, S. L., Campbell, C. P., Hindle, N., Price, J., & Scasny, R. (1998). Editing a web site: Extending the levels of edit. *IEEE Transactions on Professional Communication, 41*(1), 47–57.

Baehr, C. (2016). Certified professional technical communicator: The foundation exam and its nine areas of competency. *Intercom, 63*(1), 10–11.

Bisaillon, J. (2007). Professional editing strategies used by six editors. *Written Communication, 24*(4), 295–322. doi:10.1177/0741088307305977

Boettger, R. K. (2012). Types of errors used in medical editing tests. *AMWA Journal, 27*(3), 99–104.

Boettger, R. K. (2014). The technical communication editing test: Three studies on this assessment type. *Technical Communication, 61*(4), 215–231.

Brown-Hoekstra, K. (2017). Reviewing and editing. *Intercom, 64*(3), 18–20.

Buehler, M. F. (1981). Defining terms in technical editing: The levels of edit as a model. *Technical Communication, 28*(4), 10–15.

Buehler, M. F. (2003). Situational editing: A rhetorical approach for the technical editor. *Technical Communication, 50*(4), 458–464.

Bush, D. (1981). Content editing, an opportunity for growth. *Technical Communication, 28*(4), 15–18.

Bush, D. (1991). What are editors worth? *Technical Communication, 38*(3), 386.

Bush, D. (1992). The technology of human editing. *Technical Communication, 39*(1), 115–116.

Bush, D. (1993a). Edit yourself into a job. *Technical Communication, 40*(3), 492–494.

Bush, D. (1993b). Let the authors have their words. *Technical Communication, 40*(1), 126–128.

Bush, D. (1994a). Chopping copy. *Technical Communication, 41*(2), 322.

Bush, D. (1994b). Editing technical proposals. *Technical Communication, 41*(3), 504.

Bush, D. (1994c). Grammatical arthritis. *Technical Communication, 41*(1), 125.

Bush, D. (2000). A course in content editing. *Intercom, 47*(10), 36.

Bush, D. (2001a). Another career for editors? *Intercom, 48*(4), 38.

Bush, D. (2001b). Editing effective lists. *Intercom, 48*(8), 40.

Bush, D. (2001c). Editing is magic. *Intercom, 48*(6), 39.

Bush, D. (2002a). The professional editors network. *Intercom, 49*(5), 34–35.

Bush, D. (2002b). Three types of editing. *Intercom, 49*(9), 39–41.

Buttram, S. M. (2016). Changing the rules in an old, established game: Editing in engineering-based government agencies. *Intercom, 63*(7), 10–12.

Cantella, J., & Corbin, M. (2012). Embedding the editor: Tips and techniques for editing embedded assistance. *Intercom, 59*(8), 11–15.

Charlton, M. (2013). ETC 408/508: Technical editing. *Composition Studies, 41*(1), 101–116.

Coggin, W. O., & Porter, L. R. (1993). *Editing for the technical professions.* New York, NY: Macmillan.

Collins, W. L. (2001). Editing an index. *Intercom, 48*(2), 10.

Corbin, M. (2010). The editor within the modern organization. In A. J. Murphy (Ed.), *New perspectives on technical editing* (pp. 67–83). Amityville, NY: Baywood. doi:10.2190/NPOC5

Corbin, M., Moell, P., & Boyd, M. (2002). Technical editing as quality assurance: Adding value to content. *Technical Communication, 49*(3), 286–300.

Cox, A. L. (1981). Copy editing—The final word. *Technical Communication, 28*(4), 18–20.

Cryer, M. (2012). Hedging our bets: Using politeness in editorial comments to get results. *Intercom, 59*(8), 25–26.

Cundiff, B. S. (2017). Mechanic, mediator, mentor: Interrogating common roles of technical editors. *Intercom, 64*(6), 5–7.

Cuppan, G. P., & Bernhardt, S. A. (2012). Missed opportunities in the review and revision of clinical study reports. *Journal of Business and Technical Communication, 26*(2), 131–170. doi:10.1177/1050651911430624

Dayton, D. (1998). Technical editing online: The quest for transparent technology. *Journal of Technical Writing and Communication, 28*(1), 3–38. doi:10.2190/5EM1-R1TN-MMN3-3Y6M

Dayton, D. (2003). Electronic editing in technical communication: A survey of practices and attitudes. *Technical Communication, 50*(2), 192–205.

Dayton, D. (2004a). Electronic editing in technical communication: A model of user-centered technology adoption. *Technical Communication, 51*(2), 207–223.

Dayton, D. (2004b). Electronic editing in technical communication: The compelling logics of local contexts. *Technical Communication, 51*(1), 86–101.

Dragga, S., & Gong, G. (1989). *Editing: The design of rhetoric.* Amityville, NY: Baywood.

Dragga, S. (1996). "Is this ethical?" A survey of opinion on principles and practices of document design. *Technical Communication, 43*(3), 255–265.

Dukes, E. (1981). Some authors I have known. *Technical Communication, 28*(4), 27–31.

Eaton, A. (2010). Conducting research in technical editing. In A. J. Murphy (Ed.), *New perspectives on technical editing* (pp. 7–27). Amityville, NY: Baywood. doi:10.2190/NPOC2

Eaton, A., Brewer, P. E., Portewig, T. C., Davidson, C. R., & Portewig, C. C. (2008a). Comparing cultural perceptions of editing from the author's point of view. *Technical Communication, 55*(2), 140–166.

Eaton, A., Brewer, P. E., Portewig, T. C., Davidson, C. R., & Portewig, C. C. (2008b). Examining editing in the workplace from the author's point of view: Results of an online survey. *Technical Communication, 55*(2), 111–139.

Einsohn, A. (2011). *The copyeditor's handbook: A guide for book publishing and corporate communications* (3rd ed.). Berkeley, CA: University of California Press.

Fitzgerald, C., Stanley, J., & Lindstrom, A. (2017). Editing. Technical Communication Body of Knowledge. Retrieved from http://www.tcbok.org/wiki/editing

Flanagan, S. (2015). Intelligent content editing: A prototype theory for managing digital content. *International Journal of Sociotechnology and Knowledge Development, 7*(4), 50–54. doi:10.4018/IJSKD.2015100104

Frick, E. G., & Frick, E. A. (2009). Style manuals: The politics of selection. *Intercom, 56*(9), 9–12.

Gaitens, J. (2000). Lessons from the field: Socialization issues in writing and editing internships. *Business Communication Quarterly, 63*(1), 64–76.

Gillenwater, J. (2017). How complex is editing for plain language? *Intercom, 64*(6), 11–13.

Glick, H. (2010). Green & onscreen: Benefits of the onscreen editing method. *Intercom, 57*(4), 7–10.

Greenberg, S. (2010). When the editor disappears, does editing disappear? *Convergence: The International Journal of Research into New Media Technologies, 16*(1), 7–21. doi:10.1177/1354856509347695

Hallmark, E. (2009). How to maintain an existing corporate style guide: Suggested solutions for editors. *Intercom*, *56*(3), 21–23.

Hallmark, E., & Sowards, M. (2009). Editors and designers: 6 ideas for better collaboration. *Intercom*, *56*(9), 13–15.

Hart, G. S. (2000). Why edit on screen? *Intercom*, *47*(8), 34.

Hart, G. S. (2001). Automating your edits. *Intercom*, *48*(2), 38.

Hart, G. S. (2002a). Editing with style (sheets). *Intercom*, *49*(9), 36–37.

Hart, G. S. (2002b). Special needs: Editing tables and graphics. *Intercom*, *49*(4), 37–38.

Hart, G. S. (2002c). Substantive editing: Break it to them gently. *Intercom*, *49*(6), 34–35.

Hart, G. S. (2003). Editing tests for writers. *Intercom*, *50*(4), 12–14.

Haugen, D. (1990). Coming to terms with editing. *Research in the Teaching of English*, *24*(3), 322–333.

Haugen, D. (1991). Editors, rules, and revision research. *Technical Communication*, *38*(1), 57–64.

Hayhoe, G. F. (2010). Editing a technical journal. In A. J. Murphy (Ed.), *New perspectives on technical editing* (pp. 155–180). Amityville, NY: Baywood. doi:10.2190/NPOC9

Henkin, S. L. (2017). My top three editor roles. *Intercom*, *64*(6), 8–10.

Ketelaar, C. C. (2001). Editing on a global scale. *Intercom*, *48*(3), 17.

Kramer, R. (2003). Single source in practice: IBM's SGML toolset and the writer as technologist, problem solver, and editor. *Technical Communication*, *50*(3), 328–334.

Kreth, M. L., & Bowen, E. (2017). A descriptive survey of technical editors. *IEEE Transactions on Professional Communication*, *60*(3), 238–255. doi:10.1109/TPC. 2017.2702039

Kuhlenschmidt, S., & Mosby, C. (2001). Thinking in pixels: An editing system for electronic texts. *Journal of Technical Writing and Communication*, *31*(4), 443–444.

Lang, S., & Palmer, L. (2017). Reconceiving technical editing competencies for the 21st century: Reconciling employer needs with curricular mandates. *Technical Communication*, *64*(4), 297–309.

Lanier, C. R. (2004). Electronic editing and the author. *Technical Communication*, *51*(4), 526–536.

Lanier, C. R. (2009). Creating editorial authority through technological innovation. *Journal of Technical Writing and Communication*, *39*(4), 467–479. doi:10.2190/ TW.39.4.h

Lemanski, S. (2012). Where technical editing and journalism intersect: Stepping into unknown subject matter. *Intercom*, *59*(8), 28–31.

Mackiewicz, J. (2011). Epinions advisors as technical editors: Using politeness across levels of edit. *Journal of Business and Technical Communication*, *25*(4), 421–448. doi:10.1177/1050651911411038

Mackiewicz, J. (2014). Motivating quality: The impact of amateur editors' suggestions on user-generated content at Epinions.com. *Journal of Business and Technical Communication*, *28*(4), 419–446. doi:10.1177/1050651914535930

Mackiewicz, J., & Riley, K. (2003). The technical editor as diplomat: Linguistic strategies for balancing clarity and politeness. *Technical Communication*, *50*(1), 83–94.

MacNealy, M. S. (1999). *Strategies for empirical research in writing*. Boston, MA: Allyn and Bacon.

Malone, E. A. (2006). Learned correctors as technical editors: Specialization and collaboration in early modern European printing houses. *Journal of Business and Technical Communication*, *20*(4), 389–424. doi:10.1177/1050651906290232

Mann, M. H. (1985). How to edit the passive writer's work. *Technical Communication, 32*(3), 14–15.

Masse, R. E. (1985). Theory and practice of editing processes in technical communication. *IEEE Transactions on Professional Communication, 28*(1), 34–42.

Maylath, B. (2011). Editing for global contexts. In C. D. Rude & A. Eaton (Eds), *Technical editing* (5th ed.; pp. 300–317). Boston, MA: Longman.

McNeill, A. (2001). Technical editing 101. *Intercom, 48*(10), 10–11.

Moell, P., Corbin, M., David, M. J., Lamarche, C., & Servais, J. (2012). The evolving role of the technical editor. *Intercom, 59*(8), 6–9.

Murphy, A. J. (Ed.). (2010). *New perspectives on technical editing.* Amityville, NY: Baywood.

Murphy, A. J. (2015). How to work with your editor. *Intercom, 62*(1), 21–23.

Nadziejka, D. E. (1995). A revision of the lowest level of editing. *Technical Communication, 42*(2), 278–283.

Norman, R., & Frederick, R. A. (2000). Integrating technical editing students into a multidisciplinary engineering project. *Technical Communication Quarterly, 9*(2), 163–189. doi:10.1080/10572250009364692

Norton, S. (2009). *Developmental editing: A handbook for freelancers, authors, and publishers.* Chicago, IL: University of Chicago Press.

Osborne, H. F. (1981). Intuition, integrity, and the decline of editing. *Technical Communication, 28*(4), 21–26.

Plunka, G. A. (1988). The editor's nightmare: Formatting lists within a text. *Technical Communication, 35*(1), 37, 42–44.

Radella, M. J. (2000). Graphic electronic editing. *Intercom, 47*(9), 19.

Rogers, E. M. (1962). *Diffusion of innovations* (1st ed.). New York, NY: Free Press.

Rogers, E. M. (1971). *Diffusion of innovations* (2nd ed.). New York, NY: Free Press.

Rude, C. D. (2010). The teaching of technical editing. In A. J. Murphy (Ed.), *New perspectives on technical editing* (pp. 51–65). Amityville, NY: Baywood. doi:10.2190/NPOC4

Rude, C. D., & Eaton, A. (2011). *Technical editing* (5th ed.). Boston, MA: Longman.

Schrank, K. (2013). Using editing checklists for more efficient editing. *AMWA Journal, 28*(4), 164–166.

Seagren, R., & Gardash, L. (2012). Talking about an evolution: Improving the user interface. *Intercom, 59*(8), 16–21.

Shimberg, H. L. (1981a). Editing authors' style—A few guidelines. *Technical Communication, 28*(4), 31–35.

Shimberg, H. L. (1981b). Special issue on technical editing: Introduction. *Technical Communication, 28*(4), 4.

Soderston, C. (1985). The usability edit: A new level. *Technical Communication, 32*(1), 16–18.

Strimling, Y., & Corbin, M. (2009). Editing modular documentation: Some best practices. *Intercom, 56*(5), 28–31.

Swarts, J. (2010). Recycled writing: Assembling actor networks from reusable content. *Journal of Business and Technical Communication, 24*(2), 127–163. doi:10.1177/1050651909353307

Tarutz, J. A. (1992). *Technical editing: The practical guide for editors and writers.* New York, NY: Basic Books.

Thomas, S. (2009). More than grammar: Expectations of technical editing new hires. *Intercom, 56*(5), 26–27.

Thompson, I. K., & Rothschild, J. M. (1995). Stories of three editors: A qualitative study of editing in the workplace. *Journal of Business and Technical Communication, 9*(2), 139–169.

Van Buren, R., & Buehler, M. F. (1980). *The levels of edit* (2nd ed.). JPL 80-1. Pasadena, CA: Jet Propulsion Laboratory, California Institute of Technology.

Warren, T. L. (2010). History and trends in technical editing. In A. J. Murphy (Ed.), *New perspectives on technical editing* (pp. 29–49). Amityville, NY: Baywood. doi:10.2190/NPOC3

Weber, J. H. (2010). Copyediting and beyond. In A. J. Murphy (Ed.), *New perspectives on technical editing* (pp. 85–105). Amityville, NY: Baywood. doi:10.2190/NPOC6

White, A. M. (2012). Editing non-native English. *Intercom, 59*(8), 22–24.

Wilde, E., Corbin, M., Jenkins, J., & Rouiller, S. (2006). Defining a quality system: Nine characteristics of quality and the editing for quality process. *Technical Communication, 53*(4), 439–446.

Williams, J. D. (2003).The implications of single sourcing for technical communicators. *Technical Communication, 50*(3), 321–327.

Zook, L. (1981). Editing and the editor: Views and values. *Technical Communication, 28*(4), 5–9.

# 3

# "HOW DOES THAT MAKE YOU FEEL?"

## The Psychological Dimensions of Editorial Comments

*Ryan K. Boettger*

## Chapter Takeaways

- Describes the four psychological dimensions that underlie editorial comments and the language markers related to sentiment, social cognition, and social order.
- Investigates how editors' and authors' gender, college major, and native speaker status informs the sentiment of editorial comments.
- Provides recommendations for improving the sentiment of editorial comments, thereby strengthening future author–editor collaborations.

## Introduction

Technical editing is the most underdeveloped subfield of technical communication because unverified common knowledge often dictates its best practices. A substantial portion of this literature was written by practicing editors whose recommendations were often informed by personal experience rather than data analysis. Eaton (2010) described this body of knowledge as a collection of "cup of coffee articles" because the recommendations were useful, but only as useful as having a cup of coffee with someone and chatting about an experience (p. 9). As a result, we bring incomplete information into our technical editing classrooms and, worse, the information we can provide is not generalizable and does not necessarily inspire students to pursue the profession.

For example, Eaton and her colleagues (2008a, 2008b) explored the author–editor relationship, which previous literature often posited as adversarial, tense, and full of dissatisfaction. In fact, Eaton et al.'s survey revealed that most authors characterized their experiences with editors as useful and positive.

The results also indicated how authors typically defined editing, their preferred editing modes, and commenting structures—information that strengthens author–editor collaboration and provides concrete best practices to discuss with our students.

This chapter addresses one aspect of the author–editor relationship, specifically the psychological dimensions that underlie the comments that editors provide for their authors. Previous literature has explored obliging comment structures, but no study has yet addressed how these comments, in their totality, reflect sentiment, cognition, and social order. Understanding how these dynamics relate offers insight into how editors can motivate a more successful revision from their authors. For this study, I analyzed a 41,146-word corpus of graduate student editorial comments on four sentiment variables: analytical thinking, clout, authenticity, and emotional tone. A regression analysis also indicated whether editors varied how they delivered these comments based on their own or their authors' gender, college major, and native speaker status.

## Literature Review

The author–editor relationship is arguably the most covered topic in technical editing literature—both as lore and empirically—because it presents some of the most intriguing hierarchical dynamics in communication. In this section, I briefly describe how these dynamics interact (and sometimes compete) based on editorial objectives and the social variables of editors and authors. One strategy for improving author–editor collaboration is constructing effective comments. Therefore, I next describe the major empirical studies that have explored comment structures and suggest that also examining the psychological dimensions behind these comments can produce more informed best practices for teaching technical editing.

On a foundational level, technical editors are in demand because they enjoy "paying rigorous attention to nuance and detail, finessing the intricacies of language, and negotiating with authors" (Murphy, 2010, pp. 1–2). Authors typically report useful editing experiences, but many of them seek this assistance because it is mandated by their company (Eaton et al., 2008b). Editors are tasked with projecting professional competence, but not in ways that unnecessarily challenge the authors' writing style (and patience). In the end, authors possess final editorial control, and editors' effectiveness hinges on having their suggestions accepted. Negative editing experiences are perhaps not as common as the literature suggests, but authors complain about unneeded or excessive editing, changing the meaning of the document, and an excessive amount of time necessary to review the edited documents (Lanier, 2004). One editor's attention to detail and finessing might be one author's definition of an excessive, time-consuming process.

The dynamics that inform the author–editor relationship become further complicated when considering social variables, including gender, academic major, and native speaker status. Linguistic research has indicated that social variables correlate with syntactic choices (Adamson, 1992; Fries, 1940). Specifically, females often choose more formal registers than male speakers (Finegan & Biber, 2001) as well as use more emphatics (e.g., *really, real, such, so*) when communicating with other females (Biber, Conrad, & Reppen, 1998). Similar trends have been reported in technical communication research. For example, female technical writing students typically attend the demonstrative *this* with a noun phrase, thus choosing the more formal register (Boettger & Wulff, 2014). Female students also use more adverbs in their technical writing than males as well as associate with adverbs common to conversation, including *really, too*, and *maybe* (Boettger & Wulff, under review). Additionally, both studies reported that academic major and native speaker status influenced some of the writers' language choices, meriting similar exploration within technical editing. Survey results from Eaton et al. (2008a) again provide the best information on how social variables impact the editing process. Female and male authors as well as native and nonnative English speakers had different opinions on the most obliging commenting structure. Native English speakers were also more likely to accept proofreading and copyediting suggestions than nonnative speakers. However, as Eaton et al. point out, one of the most significant findings from their survey was that social variables did not impact editorial communication as much as perhaps presumed.

Learning—let alone conceptualizing—all these dynamics can be challenging to illustrate in the technical editing classroom. When reviewing literature for this chapter, I discovered an abundance of professional characteristics that effective technical editors "must" possess: editors must always be confident, assured, diplomatic, patient, polite, neutral, and objective. They must also account for authors' reactions to their suggestions, so they must be modest, flexible, willing to compromise, empathetic, sensitive, pleasant, and even cheery (e.g., Bostian, 1986; Dahm, 1998; Hart, 2004, 2010; Mackiewicz & Riley, 2003). It is a daunting list of requirements to share with students and one that is more derived from subjective interpretation than data-driven inquiry.

One strategy for developing these professional characteristics, and, thus, for better understanding the dynamics of the author–editor relationship, is learning to write constructive comments. I define *constructive* as comments that encapsulate both editorial expertise and empathy, leading the author to accept those comments, and, as a result, improving the overall effectiveness of the document. Again, the editing literature on commenting strategies is founded in the anecdotal, but recent research has filled this gap with more generalizable findings.

Mackiewicz and Riley (2003) began the conversation with eight linguistic strategies for balancing clarity and politeness in editorial comments. These strategies included both direct and indirect language features and accounted for many of the aspirational characteristics mentioned earlier. The most comprehensive assessment of commenting structures was conducted by Eaton and her colleagues (2008a, 2008b). The researchers assessed Mackiewicz and Riley's recommendations on over 400 authors and concluded that only two of the eight strategies performed as predicted. In fact, authors ranked a structure that Eaton et al. added as the second most obliging structure; this comment type begins with a payoff statement followed by the suggestion. More recently, Albers and Marsella (2011) analyzed the quality of 132 comprehensive comments from undergraduate technical editing students. They noted that student editors grasped paragraph-level edits and often delivered their comments with a direct style; however, they struggled with producing global-level edits related to the documents' overall structure and organization.

As technical editing research continues to empirically explore the structure of effective comments, it is important to also investigate the sentiment, cognitive processing, and social positioning that underlie these comments. Understanding linguistic structures in tandem with psychological dimensions can produce more generalizable strategies for constructing editorial comments.

This chapter begins this exploration by subjecting a corpus of student editorial comments to a sentiment analysis. Sentiment analysis is the automatic extraction of semantic information related to human feelings and opinions (Pang & Lee, 2008). Researchers use sentiment analyses to better understand how emotions, feelings, and opinions influence cognition, economic choices, learner engagement, and political affiliations (Crossley, Kyle, & McNamara, 2017).

Technical editing teachers and students can benefit from the results of sentiment analyses in many ways. First, the computerized techniques used to assess sentiment are derived by analyzing texts on multiple language markers. For example, function words (e.g., pronouns, articles, prepositions) convey information about psychological characteristics such as thinking style, honesty, and social status (Drouin, Boyd, Hancock, & James, 2017; Kacewicz, Pennebaker, Davis, Jeon, & Graesser, 2014; Newman, Pennebaker, Berry, & Richards, 2003). Understanding how language markers with little lexical meaning could signal sentiment better informs the teaching of grammar as well as the strategies developing editors use in their commenting structures. Next, sentiment analyses account for the abundance of characteristics associated with effective technical editors, namely an editor's ability to display professional competence while also protecting their authors' ego.

The sentiment variables discussed in this chapter were derived from several language categories by the developers of the Linguistic Inquiry and Word Count (LIWC). These four variables include analytical thinking, clout, authenticity, and emotional tone.

- *Analytical Thinking* encompasses the level of formality and logic a writer displays in a text (Pennebaker, Chung, Frazee, Lavergne, & Beaver, 2014).
- *Clout* accounts for the expertise and confidence that writers project (Kacewicz et al., 2014).

- *Authenticity* measures honesty and the level of disclosure (Newman et al., 2003).
- *Emotional Tone* measures the positive and negative emotions in a text (Cohn, Mehl, & Pennebaker, 2004).

Analytical thinking and clout may be particularly relevant variables to editorial comments because they reflect the hierarchical dynamics synonymous with author–editor collaborations. Editors need to exhibit clout with their authors, but they must also acknowledge authors' ownership of the text. Similarly, editors need to exhibit a high degree of analytical thinking to persuade authors to accept those comments. Authenticity and emotional tone might also provide insight into other relational dynamics, such as the openness and empathy that editors use when delivering their comments. Finally, the interactions among these four variables might also motivate inquiry into other technical texts. For example, technical communicators have hypothesized that texts with more emotional tone than analytical thinking would display a greater proportion of emotional rather than analytical word choices (Campbell, Naidoo, & Campbell, 2017). This amount of emotion could mark an inexperienced technical communicator, who cannot effectively deliver information to the intended audience.

I designed this study around the following research questions:

**RQ1** What are the psychological dimensions that underlie editorial comments? Specifically, how do the comments reflect analytical thinking, clout, authenticity and emotional tone?

**RQ2** How do social variables (gender, major, native speaker status) influence how editors deliver their comments?

## Methods

In this section, I describe the sample as well as the measures used to explore the editorial comments.

### Sample

Data were collected from 37 student editors who were enrolled across five sections of a graduate-level technical editing course. All student editors were majors in professional and technical communication. Twenty-three of the editors were female and 14 were male. Thirty-three of the editors were native English speakers, and four were nonnative speakers.

The corpus included the comprehensive comments and author cover letters for an assignment on editing job materials. The corpus included 41,146 words and 3,477 unique words. When examined by text type, the editorial comments corpus was 21,598 words and the letter corpus was 19,548 words. For this assignment, editors identified an author who needed a set of job materials for an upcoming employment opportunity. A set of job materials was either a job letter and resume or a personal statement and resume. All materials were edited in Microsoft Word with the track changes and commenting features. Editors submitted a copy of their edited materials as well as a cover letter addressed to the author that outlined the major editorial suggestions. Editors also submitted the targeted job posting as well as the author's original materials. Since editors selected their own authors for this assignment, the existence of these authors and their original job materials were verified by the instructor.

Student editors followed the suggested graduate syllabus included in the instructor manual for Rude and Eaton (2011). Editors were at midsemester when they submitted their job materials assignment and had previously completed an intensive unit on copyediting. In preparation for this assignment, editors were assigned readings related to genre (Charney, Rayman, & Ferreira-Buckley, 1992; Eaton, 2009) and editorial commenting strategies (Eaton et al., 2008a; Locker, 2003; Mackiewicz & Riley, 2003).

Further, the editors worked with 37 different authors (23 females, 14 males). Despite the equal number of female and male editors and authors, 13 female editors worked with female authors, 10 female editors with male authors, 10 male editors with female authors, and four male editors with male authors. Twenty-eight of the authors had a STEM educational background and nine had a humanities background. Thirty-two of the authors were native speakers of English, and five were nonnative speakers. Editorial comments were collected with institutional IRB approval.

## Measures

Data were subjected to a sentiment analysis via the Linguistic Inquiry and Word Count (LIWC) in addition to a regression analysis.

The LIWC is a sentiment analysis tool that analyzes texts on 80 different categories (Pennebaker, Booth, Boyd, & Francis, 2015). The proprietary LIWC dictionary contains almost 4,500 words, which were compiled from previous dictionaries and then assessed and refined by independent researchers (Pennebaker, Boyd, Jordan, & Blackburn, 2015). The latest edition of LIWC includes four summary variables based on algorithms the creators developed through their own research: analytical thinking (Pennebaker et al., 2014), clout (Kacewicz et al., 2014), authenticity (Newman et al., 2003), and emotional tone (Cohn et al., 2004). The algorithms for these summary variables are also proprietary and rescaled to reflect a 100-point scale (ranging from 0 to 100). The results section focuses on these four summary variables but additional output from LIWC is provided to better illustrate the findings.

Results from the four summary variables also served as dependent variables in the regression. The generalized regression model was $Y = \alpha + \beta 1 X + \beta 2 GRADE$ where Y is any dependent variable and X is a categorical variable, such as gender. The gender, native speaker status, and major of the editors and authors were all independent variables. [All editors were majoring in professional and technical communication and classified as non-STEM. Therefore, *Editor's Major* was not a variable explored in this study.] In addition, *Text Type* (the cover letter to the author, comprehensive comments) and assignment *Grade* were analyzed. The latter variable was included to provide a quality measure to the study. On average, editors earned an 86.03% on the assignment ($SD = 5.44$). The level of significance for this study was $p \leq .05$.

## Results

Overall, the editorial comments scored the highest percentage in *Clout* (87.05%), *Emotional Tone* (77.53%), *Analytical Thinking* (67.52%), and *Authenticity* (27.00%). Results from the regression indicated that the independent variables explained 31.6% of *Emotional Tone* (Table 3.1). Two- and three-way interactions were also identified. The following describes the results in descending order of the LIWC summary scores followed by the related regression results.

### Clout

The editorial comments averaged 87.05% in *Clout* ($SD = 11.84$, median $= 89.84$). This summary variable considers the social status, confidence, or leadership that writers display (Pennebaker, Booth, Boyd, & Francis, 2015). Technical editors use clout to project their professional competence as well as to establish trust with their authors.

Language markers associated with clout include personal pronouns, specifically how often writers reference themselves in comparison with others. Related research has demonstrated that communicators in high-status positions prefer to use second- and third-person pronouns, suggesting these leaders are more focused on group success than individual needs (Kacewicz et al., 2014).

**TABLE 3.1** LIWC summary variables sorted by adjusted R-squared, F test, and p-value

|  | Adj R² | F | p-value |
| --- | --- | --- | --- |
| Tone | 0.316 | 2.296 | .006 |
| Clout | 0.105 | 1.331 | .194 |
| Analytic | −0.097 | 0.751 | .782 |
| Authentic | −0.159 | 0.615 | .908 |

On average, 15.25% of the words in the editorial comments corpus were pronouns, 10.45% of which were personal pronouns ($SD = 2.33$). As shown in Table 3.2, technical editors preferred to deliver their suggestions to authors in second-person ($M = 6.64, SD = 2.20$).

The following examples illustrate some editorial strategies used to establish clout. The cover letter excerpt in [a] showed how one editor balanced her use of first- and second-person pronouns. She introduced her edits by highlighting details the author briefly mentioned in the job materials as well as by recalling a previous conversation with the author. The editor attributed text ownership to the author (*your resume*), but she used first-person and strong verbs (*think, include, reformat*) to deliver suggestions and convey edits.

[a]  You also noted briefly that you wrote a grant for Kid Zone that led to $7,000 in funding, which I think you should re-emphasize in an "Accomplishments" section in your resume. I remember you told me once that you had written numerous grants for Tim that were all funded. You should include these in that section as well. I also reformatted your resume so that your resume's "guts" are more offset from the section headings, which helps make better use of white space.

Similarly, the editor in [b] justified her word choice change by referencing language she found on the targeted company's website. The editor used third-person (*their website, their ideals*) to reinforce the intended audience and rhetorical situation.

[b]  The website highlights the company's policy to deliver excellent customer service. By using the wording from their website, you display to Movie Tavern that you not only share their ideals, but also have done your research on the company.

Finally, the regression results identified a three-way interaction among *Clout* and *Editor's Gender, Author's Gender,* and *Author's Major.* Clout decreased

TABLE 3.2 Personal pronouns analyzed in LIWC and rate of use in the editorial comments

| Personal Pronoun | Rate of Use % (SD) |
| --- | --- |
| I | 2.56 (1.68) |
| we | 0.13 (0.28) |
| you | 6.64 (2.20) |
| she/he | 0.09 (0.38) |
| they | 0.73 (0.57) |

when male editors communicated with male authors who had STEM educational backgrounds ($p = .03$). However, this finding must be heavily hedged as only one male editor worked with a male STEM author. This editor's letter scored a 37.86% in *Clout*, the lowest score in this summary variable.

### Emotional Tone

Technical editors use emotional tone to show empathy for the author's concerns or to acknowledge their editorial expectations. The editorial comments corpus averaged 77.53% in emotional tone ($SD = 15.68$, median = 82.27). This summary variable considers the positive and negative emotions in the text. A high percentage of emotional tone suggests a more positive, upbeat style, while an average around 50% suggests a lack of emotionality or ambivalence (Pennebaker, Booth, et al., 2015).

Language markers used to convey emotional tone include words related to social orientation and cognitive processes (Cohn et al., 2004). On average, 3.43% of the words in the editorial comments corpus were associated with positive emotions compared with 0.31% that were associated with negative emotions. In the example below [c], the editor began her letter by establishing goodwill with the author. The inclusion of positive language (e.g., *thank, you, we, opportunity, positive, happy, excited*) established a rapport with the author, but, more important, buffered the discussion of the actual editorial comments.

[c]  Thank you for the opportunity to edit your job materials. We both had positive experiences at Houston Catholic schools during our primary and secondary education, and I'm excited to see that you want to give back to the communities that shaped us from an early age. I'm also happy to see you're searching for jobs in Houston; I'm also considering moving back to Houston after finishing my Master's, so I'm excited about the possibility of living in the same city.

Editors also established emotional tone by using language to express cognitive processes. This type of language demonstrates concern for the organization and intellectual understanding of the issues addressed in the writing (Cohn et al., 2004). On average, 13.98% of the words in the editorial comments related to cognitive processes ($SD = 2.88$), and examples [d–h] illustrate how editors showed insight (*link, feel, find, think*); cause (*how, because, makes*); differentiation (*if, or, but*); and tentativeness (*hopefully, some, else*).

[d]  You might want to mention what teams you led and how you communicated effectively. Link this to your college projects.
[e]  Deleted because I've condensed the information in the second bullet. It shortens the list and makes the information easier to read.

[f]  I think this is an appropriate level of directness. If you feel it's too strong, we can find a way to ease off a bit.

[g]  I couldn't find the name of someone in HR *or* a hiring manager, *but if* you know the name, you should use it.

[h]  Hopefully some of this helps! Let me know if there is anything else I can help with, and good luck with your job search.

The regression results indicated that the independent variables explained 31.6% of *Emotional Tone*. An interaction between *Emotional Tone* and *Text Type* was also identified. In other words, the relationship between these two variables influences the statistical significance. Editors presented a significantly higher emotional tone in their author letters than in their in-text comments ($p = .001$). Examples [a] and [b] demonstrate strategies editors used in the author letters to achieve this tone.

## Analytical Thinking

The editorial comments averaged a 67.52% in *Analytical Thinking* ($SD = 15.32$, median = 68.88). This summary variable considers the formal, logical, and hierarchical thinking that writers display (Pennebaker, Booth, et al., 2015). In addition to delivering comments with authority and empathy, technical editors must understand how to synthesize their comments in ways that are logical and clear to authors.

Language markers associated with analytical thinking include function words, such as articles, prepositions, pronouns, auxiliary verbs, adverbs, conjunctions, and negations. Function words typically account for 55%–59% of written and speech communication and provide grammatical structure rather than lexical meaning to a sentence (Rochon, Saffran, Berndt, & Schwartz, 2000). However, relevant research has found that function words are often reliable markers of psychological states (Newman, Jones, & Ritter, 2016; Pennebaker et al., 2014).

On average, 51.18% of the words in the editorial corpus were function words ($SD = 3.71$), slightly outside the typical range. Table 3.3 lists the eight function word types analyzed by LIWC, their rate of use in the corpus of editorial comments, and their rate of use in a corpus of over 50,000 college admissions essays. The latter corpus was analyzed by Pennebaker et al. (2014) and serves as a comparable for the present study. For example, these researchers found that categorical language markers, including articles and prepositions, were associated with formal and precise descriptions. Articles have been associated with concrete and formal writing (Biber, 1991) and prepositions with cognitive complexity (Pennebaker & King, 1999). In contrast, impersonal pronouns, conjunctions, and adverbs have been associated with texts that are more narrative and personal in style; their use also appears to mark lower performing writers in academic settings (Pennebaker et al., 2014).

TABLE 3.3 Function word types, examples, rate of use in editorial comments corpus, and rate of use in college admissions essays corpus

| Word Types | Examples | Editorial Comments Rate of Use % (SD) | Admissions Essays Rate of Use % (SD) |
|---|---|---|---|
| Articles | *a, an, the* | 7.31 (1.91) | 6.83 (1.30) |
| Prepositions | *all, below, much* | 13.16 (1.52) | 14.71 (1.41) |
| Personal Pronouns | *I, us, you, hers, they* | 10.14 (2.33) | 10.88 (2.05) |
| Impersonal Pronouns | *it, this, anything* | 5.82 (2.18) | 5.03 (1.38) |
| Auxiliary verbs | *are, did, have* | 7.40 (1.83) | 8.25 (1.72) |
| Adverbs | *even, just, usually* | 3.96 (1.38) | 3.90 (1.04) |
| Conjunctions | *and, so, until* | 5.92 (1.32) | 6.41 (1.02) |
| Negations | *no, never, not* | 0.73 (0.64) | 1.04 (0.49) |

The editorial comments included more articles than the admissions essays (7.31% compared with 6.83%) but fewer prepositions (13.16%, 14.71%). The remaining function words were relatively balanced between the corpora; however, the editorial comments contained slightly more impersonal pronouns (5.82% compared with 5.03%).

The following examples of articles and prepositions illustrate ways editors used formal and precise descriptions of objects, events, goals, and plans. In the cover letter excerpt [i], the editor used articles (*the, an*) to describe differences between the versions of the edited document; the version with all the Microsoft Word changes tracked, and the version with all those changes accepted. Another editor [j] used articles to instruct the author to include a phrase in her cover letter that echoes language used in the job posting. Further, in another cover letter excerpt [k], the editor used prepositions (*of, about, to, than*) to encourage the author to modify how she presented her performance testimonials in her resume. In the final example [l], the editor used prepositions to justify his edit (*since*) and to denote content placement (*above*).

[i]  The first part is an edited version of the original document. The second part is the edited document with all changes and comments listed using Microsoft Word's Review function. This allows you to review each change and comment individually.

[j]  Add a phrase to convey that you are an active, modern artist. This allows you to echo a phrase from the job description that asks for "substantial knowledge of modern and contemporary art."

[k]  Also, I think your approach of including positive comments about your work is of value, but you might consider placing them on a separate attachment devoted solely to these quotes, rather than inserting them directly in the resume. Their inclusion in the resume makes it a bit busy and, at times, difficult to follow.

[l]  The inclusion of the email seems redundant since it is listed above.

The regression identified an interaction between *Analytical Thinking* and *Grade* ($p = .010$). For every 1-point increase in grade, analytical thinking scores were expected to decrease by 1.146 +/− 0.425 (Figure 3.1). This is an interesting finding as Pennebaker et al. (2014) found that categorical language was consistently linked with better academic performance, whereas dynamic language was not. Further, the researchers found these effects were consistent across academic disciplines.

### Authenticity

The editorial comments averaged 27.00% in *Authenticity* ($SD = 17.82$, median = 21.99). This summary variable considers how writers reveal themselves in an authentic or honest way. A lower number suggests a more guarded, distant form of discourse (Pennebaker, Booth, et al., 2015). Technical editors need to be authentic in their comments but also to maintain some distance, in part to demonstrate detachment from the document.

Relevant research on the summary variable has primarily focused on language markers that distinguish true from false stories as well as in emotional text types like health narratives and poetry. Therefore, the language markers associated with low authenticity include fewer self-references, more negative emotion words, and fewer markers of cognitive complexity (Newman et al., 2003). The editorial comments examined in this study were not compared with another corpus, so these types of findings may not directly inform author–editor collaborations.

For example, technical editors included fewer self-references, but their strong use of second-person/you-inclusive language likely contributed to their clout score. Similarly, the editorial comments included a small amount of negative emotion words (0.31%); however, the words the LIWC dictionary classified as

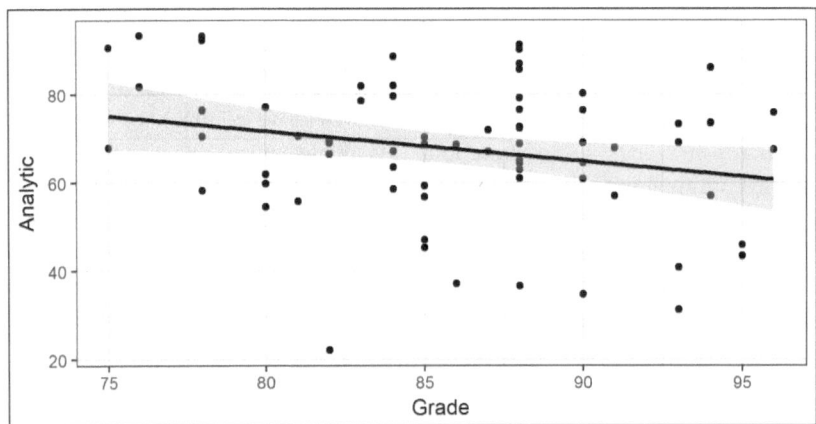

FIGURE 3.1 Interaction between Analytical Thinking and Assignment Grade.

negative depended on context. The word *tense* appears in the editorial comments corpus 14 times, but it is never associated with a negative situation. In fact, *tense* collocated with words like *past, present,* and *verb* [m]. Similarly, the editor in [n] uses the word *cut* to express concern that the author's contact information will get cut off during the printing process if it is in the Microsoft Word document header instead of within the text.

[m] Importantly, make sure you use past tense for jobs you no longer work at. Many of the older jobs flip between present and past tense.

[n] Also, your contact information was taken out of the header and placed in the body of the document. This ensures that the information does not accidently get *cut* off when printed.

The regression identified a significant two-way interaction among *Authenticity* and *Author's Gender* and *Author's Major* ($p = .043$). Authenticity appeared to increase if the author was male and had a STEM educational background. The editors worked with six different male STEM authors.

## Discussion

The student editors in this study delivered comments that showed expertise and empathy. Overall, their comments included high percentages of clout, emotional tone and analytical thinking. The following section discusses the results of this study.

Editors' balance of personal pronouns reflected their attempts to establish clout. Specifically, the comments focused substantially more on the author (*you, your*) and intended readers (*they, their*) than the editor (*I, my*). An increased use of pronouns like *we* and *you* reflect that high-status individuals are more collectively oriented than other-oriented (Kacewicz et al., 2014). While authors make the final decision to accept or reject editorial suggestions, the editors must first establish professional competence. They establish this competence by focusing their attention outward, toward their authors. The language features that inform LIWC's clout variable encompass many of the recommendations for creating a you-attitude, a writing style that considers communication from the reader's point of view (Locker, 2003). Students will likely be familiar with techniques for establishing goodwill; however, having them explore how they apply personal pronouns in their own writing assignments (and even in their personal emails and text messages) can make abstract concepts like balancing power dynamics more concrete. Examples [a]–[b] model ways that editors established clout with their authors.

The editorial comments also contained a high percentage of emotional tone, suggesting a more positive writing style. Establishing a rapport with authors becomes increasingly important as editors are tasked with delivering *constructive* criticism (but criticism, nonetheless). Student editors in this study used positive language related to insight, cause, and differentiation that delivered the comments with authority but also acknowledged authors' ownership of their job materials. The regression also showed the author cover letters gave editors a significant opportunity to use an emotional tone. In their cover letters, editors thanked authors for the chance to review their work as well as outlining how to approach the revision process. The letters functioned as a rhetorical buffer that acknowledged the anxiety synonymous with opening electronic files marked with red tracked changes. However, as with any emotional appeal, editors must be mindful of using a tone that could conceal any legitimate structural and organizational issues with the document. Authors are typically more concerned with understanding the comment rather than the structure used to convey that comment (Eaton et al., 2008a, 2008b; Mackiewicz & Riley, 2003). Editors must be aware that too much emotional tone can diminish their perceived clout and analytical thinking. Examples [c]-[h] model ways that editors established emotional tone with their authors.

Editors' use of categorical language (articles, prepositions) also accounted for a high percentage of analytical thinking. Articles are associated with concrete and formal writing and prepositions are used to link ideas. On the other hand, dynamic language is associated with impersonal pronouns, conjunctions, and adverbs as well as with lower academic performance (Pennebaker et al., 2014). Interestingly, the regression indicated that editors who earned the lowest assignment grades often displayed higher analytical thinking. While this finding contrasted with the Pennebaker et al. (2014) study of college admissions essays, a closer examination found that these student editors also scored below the average on clout or emotional tone. Examples [i]-[1] model ways that editors established analytical thinking with their authors.

In contrast, the editorial comments included a low percentage of authenticity, which considers how writers reveal themselves in an authentic or honest way. As mentioned earlier, the relevant research that informed this LIWC summary variable was conducted to distinguish true from false stories (Newman et al., 2003) as well as measure the level of disclosure in emotional narratives (Boals & Klein, 2005). Therefore, the language markers associated with low authenticity include fewer self-references, more negative emotion words, and fewer markers of cognitive complexity. However, the low frequency of editorial self-references was supplemented with a high frequency of references to authors, resulting in the high percentage of clout. Similarly, words classified as negative in LIWC did not typically connect to any negative sentiments within editorial comments. For example, *tense* was associated with verb tense rather than a heightened situation. Anecdotally, I explored the *Authenticity* variable in a variety of other technical text types. The LIWC reported a score of 21.69% on a corpus of critical reviews and 14.93% on a corpus of white papers. These observations further suggest that this LIWC variable might not be suitable for studying technical writing. In fact, consistently low scores in *Authenticity* might instead suggest that effective technical writing contains an appropriate level of professional distance with its readers. Examples [m]–[n] model ways that editors established authenticity with their authors.

The regression identified few significant interactions related to the psychological dimensions and the gender, major, and native speaker status of the editors and authors. The insignificance here, however, might actually be of significance to technical editing students and teachers. Editors did not adjust how they constructed their comments based on these social variables. Eaton et al. (2008a) made similar observations as well. However, other variables might inform these comments and should be explored in future studies. For example, the fact that authors identified (and therefore knew) their author for this assignment could have resulted in the high percentage of emotional tone.

## Pedagogical Applications

The results of the study have implications for teaching students the psychological dimensions that underlie editorial comments. Understanding how editors apply these dimensions could further improve author–editor collaborations.

First, it is important to highlight that students who participated in this study were responding to an authentic situation rather than a scenario. Allowing students to interact with actual authors allows them to apply the concepts they read about and discuss in class. Students identified their own author for this assignment; however, this familiarity did not appear to decrease engagement level. In fact, the high percentage of clout suggests the editors used the assignment as an opportunity to demonstrate their developing expertise.

When introducing students to the concepts of clout, analytical thinking, authenticity, and emotional tone, begin classroom discussion on a conceptual level. The examples provided throughout this chapter can be used to illustrate these applications and motivate class discussion. Describe the four dimensions and their related language markers and then facilitate students' initial applications of these dimensions.

Once students understand how language markers convey sentiment, encourage them to explore these concepts in their own technical writing as well as in their personal emails and text messages (as a comparison). The LIWC is not a freely available software program; however, you can emphasize how function words—pronouns, articles, and adverbs— reflect cognitive processes and social order. This type of discussion can also contextualize grammar and copyediting, which, while foundational, are typically the drier instructional units for students.

You can also use technology to assist students' understanding of these psychological dimensions. As a classroom activity, ask students to calculate the Empathy Index of a piece of writing by using the find and replace feature in Microsoft Word. The Empathy Index measures reader focus (Cleland, 2014). To calculate the index, count the number of references to the target readers and subtract it from the number of references to the writer or the organization. The find and replace feature allows students to highlight and count these instances with ease. Ideally, students should end up with a positive number; the higher the number, the more reader-focused the writing. This activity can connect to a discussion of all four psychological dimensions.

Introduce your students to AntConc (Anthony, 2006) if you want to explore language markers through a more fine-grained approach. AntConc is a freely available text processing program that allows users to generate word lists, concordances, and collocates of texts. Boettger and Wulff (2014) provided a guide for integrating text processing tools into the technical editing classroom and steps on how students could create their own corpus. A class corpus of editorial comments, for example, could help students refine their editorial style over the semester. Teachers can build this corpus over several semesters to increase the generalizability of the findings.

## Conclusions

This study is not without its limitations. Sentiment analysis is one way to analyze language and any results produced by it should be scrutinized. Sentiment analysis tools are also created differently. The indices in LIWC, for example, are based on simple word counts that do not consider issues like negations or part-of-speech tags. Similarly, the algorithms for the four LIWC summary variables are proprietary, which can make analysis more difficult. New sentiment analysis tools are becoming increasingly available, and their construction and findings can benefit the field of technical communication.

The sample explored in this chapter also invites its own limitations. This study examined the editorial comments written by developing student editors (rather than professional editors) on a single assignment. Additional data could suggest how the editorial tone of students evolves over the course of the semester. Additionally, students selected their authors for this project, which could have influenced the results. Finally, the social variables explored in this study are only a few factors that could influence how editors communicate with their authors. Future studies in this area as well as studies on how authors perceive the clout, emotional tone, analytical thinking, and authenticity of the comments can improve the author–editor relationship and further advance the scholarship in technical editing.

## Pedagogical Practicalities

- Understanding the language markers that inform clout, analytical thinking, emotional tone, and authenticity can improve how technical editors deliver comments to their authors.

- The examples in this chapter are included to motivate in-class discussion about how to construct editorial comments as well as to encourage students' independent learning.

## References

Adamson, H. D. (1992). Social and processing constraints on relative clauses. *American Speech, 67*(2), 123–133. doi:10.2307/455450

Albers, M. J., & Marsella, J. F. (2011). An analysis of student comments in comprehensive editing. *Technical Communication, 58*(1), 52–67.

Anthony, L. (2006). Developing a freeware, multiplatform corpus analysis toolkit for the technical writing classroom. *IEEE Transactions on Professional Communication, 49*(3), 275–286. doi:10.1109/TPC.2006.880753

Biber, D. (1991). *Variation across speech and writing.* New York, NY: Cambridge University Press.

Biber, D., Conrad, S., & Reppen, R. (1998). *Corpus linguistics: Investigating language structure and use.* New York, NY: Cambridge University Press.

Boals, A., & Klein, K. (2005). Word use in emotional narratives about failed romantic relationships and subsequent mental health. *Journal of Language and Social Psychology, 24*(3), 252–268. doi:10.1177/0261927X05278386

Boettger, R. K., & Wulff, S. (2014). The naked truth about the naked *this*: Investigating grammatical prescriptivism in technical communication. *Technical Communication Quarterly, 23*(2), 115–140. doi:10.1080/10572252.2013.803919

Boettger, R. K., & Wulff, S. (under review). Gender effects in student technical writing— A corpus-based study.

Bostian, L. R. (1986). Working with writers. *Scholarly Publishing, 17*(2), 119–126.

Campbell, K. S., Naidoo, J. S., & Campbell, S. M. (2017). *Hard- and soft-sell marketing in white papers.* Paper presented at the Professional Communication Conference (ProComm), 2017 IEEE International, Madison, WI, July 23–26.

Charney, D., Rayman, J., & Ferreira-Buckley, L. (1992). How writing quality influences readers' judgment of resumes in business and engineering. *Journal of Business and Technical Communication, 6*(1), 38–74. doi:10.1177/1050651992006001002

Cleland, J. K. (2014). *Business writing for results: How to create a sense of urgency and increase response to all of your business communications.* New York, NY: Diversion Books.

Cohn, M. A., Mehl, M. R., & Pennebaker, J. W. (2004). Linguistic markers of psychological change surrounding September 11, 2001. *Psychological Science, 15*(10), 687–693. doi:10.1111/j.0956-7976.2004.00741.x

Crossley, S. A., Kyle, K., & McNamara, D. S. (2017). Sentiment Analysis and Social Cognition Engine (SEANCE): An automatic tool for sentiment, social cognition, and social-order analysis. *Behavior Research Methods, 49*(3), 803–821. doi:10.3758/s13428-016-0743-z

Dahm, R. (1998). *No pain, no shame editing.* Paper presented at the Annual Conference— Society for Technical Communication, Anaheim, CA, May 11–13.

Drouin, M., Boyd, R. L., Hancock, J. T., & James, A. (2017). Linguistic analysis of chat transcripts from child predator undercover sex stings. *The Journal of Forensic Psychiatry & Psychology, 28*(4),1–21. doi:10.1080/14789949.2017.1291707

Eaton, A. (2009). Applying to graduate school in technical communication. *Technical Communication, 56*(2), 149–171.

Eaton, A. (2010). Conducting research in technical editing. In A. J. Murphy (Ed.), *New perspectives on technical editing* (pp. 7–28). Amityville, NY: Baywood.

Eaton, A., Brewer, P. E., Portewig, T. C., & Davidson, C. R. (2008a). Comparing cultural perceptions of editing from the author's point of view. *Technical Communication, 55*(2), 140–166.

Eaton, A., Brewer, P. E., Portewig, T. C., & Davidson, C. R. (2008b). Examining editing in the workplace from the author's point of view: Results of an online survey. *Technical Communication, 55*(2), 111–139.

Finegan, E., & Biber, D. (2001). Register variation and social dialect variation: The register axiom. In P. Eckert & J. R. Rickford (Eds), *Style and sociolinguistic variation* (pp. 235–267). Cambridge, UK: Cambridge University Press.

Fries, C. C. (1940). *American English grammar.* New York, NY: Appleton-Century Company.

Hart, G. (2004). Helping authors work with you. *Intercom, 51*(7), 38–39.

Hart, G. (2010). The editor and the electronic word: Onscreen editing as a tool for efficiency and communication with authors. In A. J. Murphy (Ed.), *New perspectives on technical editing* (pp. 107–126). Amityville, NY: Baywood.

Kacewicz, E., Pennebaker, J. W., Davis, M., Jeon, M., & Graesser, A. C. (2014). Pronoun use reflects standings in social hierarchies. *Journal of Language and Social Psychology, 33*(2), 125–143. doi:10.1177/0261927X13502654

Lanier, C. R. (2004). Electronic editing and the author. *Technical Communication, 51*(4), 526–536.

Locker, K. O. (2003). Building goodwill. In *Business and administrative communication* (6th ed., pp. 32–55). New York, NY: McGraw-Hill.

Mackiewicz, J., & Riley, K. (2003). The technical editor as diplomat: Linguistic strategies for balancing clarity and politeness. *Technical Communication, 50*(1), 83–94.

Murphy, A. J. (2010). Introduction. In A. J. Murphy (Ed.), *New perspectives on technical editing* (pp. 1–5). Amityville, NY: Baywood.

Newman, D. S., Jones, J., & Ritter, C. (2016). Vertical peer supervision of consultation: A linguistic exploration of relational hierarchy. *The Clinical Supervisor, 35*(2), 287–304. doi:10.1080/07325223.2016.1218811

Newman, M. L., Pennebaker, J. W., Berry, D. S., & Richards, J. M. (2003). Lying words: Predicting deception from linguistic styles. *Personality and Social Psychology Bulletin, 29*(5), 665–675. doi:10.1177/0146167203029005010

Pang, B., & Lee, L. (2008). Opinion mining and sentiment analysis. *Foundations and Trends® in Information Retrieval, 2*(1–2), 1–135. doi:10.1561/1500000011

Pennebaker, J. W., Booth, R., Boyd, R., & Francis, M. (2015). Linguistic inquiry and word count: LIWC2015 operator's manual. Retrieved April 28, 2016. In Pennebaker, J. W., Booth, R., Boyd, R., & Francis, M. (2015). Linguistic inquiry and word count: LIWC 2015 [computer software]. Austin, TX: Pennebaker Conglomerates: Inc.

Pennebaker, J. W., Boyd, R. L., Jordan, K., & Blackburn, K. (2015). *The development and psychometric properties of LIWC2015.* Austin, TX: University of Texas at Austin. Retrieved from https://repositories.lib.utexas.edu/bitstream/handle/2152/31333/LIWC 2015_LanguageManual.pdf

Pennebaker, J. W., Chung, C. K., Frazee, J., Lavergne, G. M., & Beaver, D. I. (2014). When small words foretell academic success: The case of college admissions essays. *PloS one*, *9*(12), e115844. doi:10.1371/journal.pone.0115844

Pennebaker, J. W., & King, L. A. (1999). Linguistic styles: Language use as an individual difference. *Journal of Personality and Social Psychology*, *77*(6), 1296. doi:10.1037/0022-3514.77.6.1296

Rochon, E., Saffran, E. M., Berndt, R. S., & Schwartz, M. F. (2000). Quantitative analysis of aphasic sentence production: Further development and new data. *Brain and Language*, *72*(3), 193–218. doi:10.1006/brln.1999.2285

Rude, C. D., & Eaton, A. (2011). *Instructor's manual to accompany technical editing* (5th ed.). Boston, MA: Longman.

# 4

# IMAGINATION AS AGENCY

## Communities of Practice and Editing Pedagogy

*Tracy Bridgeford*

## Chapter Takeaways

- An understanding of how to use narratives as contexts for editing assignments.
- An understanding of how imagination works as a mode of belonging in a community of practice.
- An understanding of how imagination functions as a mechanism for reflection, agency, and identity in editing.

## Introduction

The pedagogical model I present in this chapter represents both an application and evolution of the communities-of-practice approach to teaching technical communication that I refined over several years of application (see Bridgeford, 2007). This pedagogical framework employs Scott Russell Sanders' science-fiction novel *Terrarium* to provide political, professional, and intellectual context and a narrative backdrop for student work. This approach grows out of the longstanding humanistic tradition sparked by Miller's (1979) landmark article, "A Humanistic Rationale for Technical Writing." For Miller, understanding "how to belong to a community" is vital for professional enculturation, for establishing membership in and accepting responsibility within a community (p. 617). This focus on ethical, rhetorical enculturation is particularly important to current practice in technical communication, especially when we consider the challenges that come with preparing new professionals to design documents through content management systems (CMSs).

For reasons I explore later in this chapter, I was at first resistant to the notion of conducting my own work in a content management system (CMS). I was trained as a humanistic scholar during the late 1990s, and was enculturated into the social scholarly perspective described by Blyler and Thralls (1992) and Thralls and Blyler (1993). When I became Information Officer for the Council for Programs in Technical and Scientific Communication (CPTSC) in 2003, I was confronted with the organization's CMS. I resisted using it to maintain the Council's online presence, forgoing it for a traditionally designed website with content composed page-by-page. Thinking back, my resistance was rooted in the connections I believed could not be compromised between a humanistic approach to design and professional responsibility, connections I felt would somehow be compromised by work in a CMS. However, further reflection suggests that I was both right and wrong. As I considered both my professional experience and my pedagogy, I discovered that although a CMS presents significant editorial and ethical issues, we can still conduct ourselves from a humanistic perspective. A description of this humanistic approach to teaching editing is one that relies on narrative ways of knowing to situate action (Bridgeford, 2004, 2007).

Because the day-to-day activities of editorial work are covered extensively by other scholars (e.g., Albers, 2000; Dayton, 2003; Kreth & Bowen, 2017; Rude & Eaton, 2011), I focus on the pedagogical framework I create in the editing classroom, and specifically on how it applies to belonging to and participating in the practices of a community. I follow that discussion with an examination of Wenger's (1998) concept of an *identity of participation*—his notion of agency—from which I emphasize imagination as a mode of belonging, and discuss the role stories play in building individual identity and confidence in a community of practice (COP). Next, I present a personal narrative about my own editing experience to demonstrate how I came to develop this pedagogical framework. I then describe how I use *Terrarium* and a repository of student writing as an imaginative context for assignments. Finally, I discuss how imagination empowers students to develop a sense of identity and belonging that helps them see themselves as professionals with the competence and authority to manage the editorial process and to make changes to content.

## Humanistic Approaches and Technical Communication

Disciplinary conceptions of community in technical communication classrooms were impacted significantly by Miller's (1979) concept of enculturation. This process of understanding "how to belong to a community" is Miller's pedagogical pivot that shifted scholarly discussions toward the social and political, changing the philosophical landscape of technical communication (p. 617). This moment continues to resonate in the discipline's ongoing work, and thus continues to affect our classrooms despite technological changes. Although technology systems have never been our primary focus in this context; interconnections between technologies and human systems have been essential to our work.

Content management systems challenge our discipline and profession in many ways, but their impact on how we establish and sustain a sense of community has not been discussed in sufficient depth. Miller's concept of community is helpful here because it demonstrates how writing and editing define not only the practices of a community but also its identity. The role of identity goes beyond simple membership to include acts of authority:

> to write [or in this case, to edit], to engage in any communication, is to participate in a community; to [edit] well is to understand the conditions of one's own participation—the concepts, values, traditions, and style which permit identification with that community and determine the success or failure of communication.
>
> (Miller, 1979, p. 617)

Without the concepts, values, traditions, and stylistic standards and expectations to inform our writing and editing, there is only technology and writing mechanics with no humanistic value.

In many ways, the novel I use to provide context in my technical editing class (which I discuss later) parallels the scholarly discourse of the 1980s by offering yet another version of the tension between instrumental and humanistic approaches to technical writing (Johnson, 1998; Miller, 1979; Moore, 1996). An instrumental approach to teaching is arhetorical, focusing on correctness and clarity (Moore, 1996). A humanistic approach, on the other hand, moves away from what Miller (1979) called the "windowpane theory of language" (p. 611) or a transparent medium with no agentive element, to consider motive and responsibility in communication action. Miller's humanistic rationale became a "defining point of text in [the dominating] discussions," that followed in the discipline (Smith, 1997), inspiring others to build further on rhetorical traditions: Sullivan's (1990) political-ethical implications of defining technical communication as practice; Johnson's (1998) user-centered design based in *praxis* and *techne*; Longo's (2000) blending of cultural studies and classical rhetoric; Johnson-Eilola's (1996) framing of technical communication as symbolic-analytical work. Each of these scholars has refined, expanded, and further promoted humanistic perspectives and approaches to scholarship and pedagogy.

Today, I find myself surprised that I must still defend a humanistic rationale for technical communication to some colleagues. This historically significant discussion, establishing knowledge that we in the discipline perhaps have come to take for granted, seems to come back around when new communication technologies emerge. Current discussions about content management systems and the writing practices they privilege have been rhetoricized by scholars such as Pullman and Gu (2008). The discipline's pedagogical attention to audience, to ethical authorship, to accountability, and to usability emphasizes the value we place on humanistic perspectives and everything they demand of designers, thinkers, and teachers. Subsequent theoretical intersections, such as social constructionism, feminism, and postmodernism, might not have gained traction without the opening Miller created.

Professionally, much of the technical communication world has already completed the transition from craft composition to single-sourced and multisourced dynamic content composition. However, the academy trails behind, both scholastically and pedagogically. Many campuses remain ill equipped even to teach single-author, linear, self-contained texts, at least technologically. Few campuses support writing through CMSs, at least in technical communication programs. Such practices are well established in computer science departments, where information science programs have taught database-driven web design for quite a while. Teachers new to CMSs need transitional pedagogies—stepping stones, if you will—to get from one composition framework to the other. To help students cross this divide, I recast some traditional, craft-based editing strategies with the help of Baker's (2013) concept of designing websites so that "every page is page one," and Wenger's (1998) notion of collective, or community practice.

In my teaching, I have valued the humanistic perspective in ways that enabled me to bring a science-fiction novel into the classroom as a context for editing and writing assignments, and to provide a gateway for narrative ways of knowing as a pedagogy of enculturation (Bridgeford, 2004). In my scholarship and professional activities, I have valued Wenger's communities-of-practice social learning theory. It has informed my pedagogy with the means for reifying knowledge, participating in the practices of a community, and building an identity through the modes of belonging, specifically engagement, alignment, and imagination (Bridgeford, 2007). Along the way, I've learned that narrative conditions provide an opening for agency (Bruner, 1990, 1991), empowering students with the ability to engage in the work of technical writers and editors.

With every technological movement, writing craft is seemingly discarded or devalued. But I prefer to think of craft as being reinvented or recontextualized with each significant technological shift. By employing narrative ways of knowing and communities of practice (COP), my pedagogical approach frames editing activities in ways that parallel the current shift to CMSs and dynamic text (text stored in a database and called forth by writers' and users' choices). It's topic-based writing that can be recast in multiple publishing forms according to need and situation. Reflected in this technological shift is the disciplinary tension between instrumental and rhetorical perspectives on technical communication highlighted in the 1980s by the scholarly debates among Miller (1979), Johnson (1998), Moore (1996, 1999), and Tebeaux (1980). Tebeaux (1980) challenged Miller's humanistic rationale, arguing that rhetorical pedagogy does not offer students what they need from a technical writing class. Tebeaux proposed instead that we "prepare [students] for the writing they will have to do in business and industry" (p. 822), which should include emphasis on mechanical correctness rather than political awareness. Despite such early (and ongoing) resistance to the notion that technical writing is rhetorical rather than instrumental, the discipline has embraced social, critical, audience-aware approaches to teaching and doing information design. Among those voices, Johnson-Eilola's (1996) "Relocating the Value of Work" refutes the premise that our academic work should serve as a "wish-list for industry" (p. 247). Every discipline bears responsibility for understanding the implications of its own workplace contributions, and thus of accepting responsibility for its actions.

I've never really understood arguments against the humanistic perspective and its rhetorical foundation, given that industry rewards critically aware, humanistic design work, regardless of the technology used to connect audiences with knowledge. Lanier (2012) and Rockley (2003) remind us that single sourcing (also called component content management) is really about people, not just technology. From one perspective, single sourcing, the writing of content at the granular level, could appear to be formulaic, because it requires writers to structure their writing in ways that make content usable across a spectrum of media, thus separating content and presentation. Some see this perspective as a return to the "positivistic" perspectives we rallied against in the 1980s (Bacha, 2009). However, as Rockley (2003) argues, structured writing, or topic-based writing, "frees [technical writers] from some of the more mundane components of writing" and provides a "way to improve the quality of their information [to] save money and time for the organization" (p. 351). Crafting content asks that writers imagine documents as coherent wholes, while designing dynamic content asks that writers think instead about topics as coherent wholes. Topic-based writing can thus stand alone regardless of the media used to publish it.

Although this change may bring frustration to writers who view their writing from the craft perspective of authorship and ownership, the value of collaborative writing, or as Rockley and Cooper (2012) call it, "true collaboration" (2012, p. 224), expands in that writers become responsible for a particular part of a product line's content, for example, instead of for an entire manual. Writers work together to create an array of content chunks that can stand alone or be joined with other content chunks. In my classroom, I work to defuse the tension between these traditional and the emerging schools of thought through narrative ways of knowing. That is, by focusing learning activities through narrative frames à la Wenger (that is, by telling and interpreting stories that serve as a contextual backdrop for action), I enact one method for bridging that gap. I create an economy of practice, one I have found useful across the context of many courses and many years. By working together, groups of editors and writers create economies of practice, that is, organic constructions of content through collaborative efforts of communities working toward the same goal.

Regardless of the technologies used by designers to execute their work, two constants remain: (1) professionals engage the world through narrative ways of knowing, and (2) community is constructed through collective imagination. Both of these things are essential to our ability to see ourselves as writers or editors in

any working context. Without imagination, we would not be able to see possibilities. We would not be able to learn, and we would not be able to join a community. Imagination is what gives us the source material upon which we reflect, and which we come to understand as practice. Its purpose is to guide us along a path to learning, to enculturation, and to belonging and participation. My own story, which I share later, demonstrates the path I think might be familiar to those that students take as they begin to imagine themselves as editors. As it has been for me, the role and work of imagination begins for my students, I think, with the craft tradition of composition and design. We observe others at work, we emulate their practice, and then we practice ourselves. This is writing and design craft regardless of the technology used at the time. My father, who was a master craftsman of woodworking and cabinetry, once told me that if you can't picture what you're doing, you won't be able to do it. My approach to teaching editing echoes his approach to woodworking and begins with imagination (sparked by a novel), situates the action (in plausible, but fictional context), and invites students to enter the text (through interpretation and action). We can always teach students the techniques of editing. What we must also do is teach them to see themselves as editors. This act requires imagination.

I begin with attention to writing economies. Specifically, I emphasize editing passages of text to distill from them their essential content, expressed with economy, and packed with meaning. Editing word by word and line by line is the most sacred tenet of the craft tradition for composing texts—an editor or editing team expressing care for content, meaning, and voice through attention to the smallest (and sometimes most esoteric) rhetorical details and strategies. This core technical communication strategy is itself the essence of effective web communication, and, it turns out, is also the essence of effective topic- and task-driven writing for content assembled through CMSs. It is a relatively easy transition for students to get from economic writing and editing to content nodes, or self-contained, coherent texts that begin and end in a single content element. This is Baker's (2013) notion of successful content development for the web, or for any other context where there is an emphasis on understanding and serving the needs, expectations, and consumption habits of contemporary readers. Once students become comfortable with the shift from editing for economy to writing for immediate coherence, they are ready to compose or edit for CMS-driven design environments.

## Communities of Practice

My narrative-based pedagogy resulted from reflection at the intersection of Miller's (1979) humanistic approaches to technical communication and social learning theory, specifically Wenger's (1998) concept of communities of practice. This framework has evolved and expanded more recently to address the challenges of composing and editing in and through CMSs. Wenger (1998, 2015) defines a community of practice (COP) as a group of "people who share a concern or a passion for something they do and learn how to do it better as they interact regularly." Members of a COP are drawn together through a joint enterprise and a social fabric that guides community members and their participation, both of which grow out of members' engagement. To participate in the practices of a community, members must develop a sense of belonging (identity), cultivate it through engagement (participation), match their interests with the community's practice (alignment), and form images of reality (imagination). As part of that participation, these characteristics give rise to a COP through members' accountability to the community. To contribute, to affect the meaning of the enterprise, members develop an *identity of participation*—Wenger's notion of competent membership—refers to the image of what constitutes competence for that community.

Wenger's (1998) COP is perhaps more conceptual than concrete. That said, technical communication's humanistic reinvention that transpired in the 1980s seeks a balance of theory and practice in its pedagogy. I like Wenger's commitment to community values, to enculturation of novice and inexperienced members through shared narratives that inform and ultimately impact daily practice. Thus, I give a nod to Wenger in framing this transitional pedagogy as *economies of practice*. So much of the practical and conceptual value of CMSs is inextricably intertwined with economy—even at the most simplistic and direct level, single-source documents save time and money by making writing economical. More broadly and more richly, CMS-driven design is built upon an undergirding economy, upon maximizing the value of content, talent, and thus ultimately of technical writers themselves. In a CMS environment, content is the most valuable element of the work, and publishing forms become more transient, more contextualized and situated, and ultimately of less durable value. Not devalued, but rather more momentary in value. The discipline's collective consideration of the rhetorical, social, and political implications of CMSs is still evolving. Although humanistic scholars are cautious to embrace economically framed or driven processes unless we can be sure they are critically vetted, we must also recognize the forces in play in contemporary workplaces, writing spaces, and even conceptual frames. Our pedagogies serve in a variety of contexts, including industrial ones that are economically motivated and measured. Perhaps part of our hesitance in adopting content management systems stems from the possibility that they render our pedagogies as less durable as well. Craft composition, and thus craft pedagogy, prides itself on the investment of time to get something just right. To think that we might invest so deeply in something of fleeting value threatens to undermine our locus of personal and professional value.

An aspect of COPs that contributes to building an identity of participation is imagination. By using their imagination, members learn to belong to a community through images they make of their and others' worlds. For Wenger, imagination "emphasizes the creative process of producing new 'images' and of generating new relations" (1998, p. 185), both of which guide the meaning of the community's practice—a meaning that is accessed first through the imagination. Without imagination, members would not be able to engage in the practice. Imagination is what enables members to "recognize . . . experience" in themselves and other community members who guide their ability to develop competence. It is, Wenger says, "imagination [that] creates a kind of community . . . to which [members] see themselves as belonging" (pp. 181–182). Without imagination, community members would find it more difficult to build an identity of participation. Imagination thus becomes a process and strategy, not a thing.

For students in my classes, this participation is framed by the social and political context of Sanders' *Terrarium. Terrarium* is a dystopian novel depicting an Earth that has become so polluted that human beings are no longer able to survive unaided in that environment. To solve the problem, they build huge, globe-like enclosures to house humanity. This novel works well in technical writing and editing courses because it involves a technological solution to a grand set of ecological and social problems, and thus situates students as agents for humanity and technology. Its apocalyptic story is close enough to current circumstances to create a plausible reality that students can share. It unifies the class within a particular context. Students interact with each other and the novel's characters, engaging in their own constructed communities of practice. Imagination's role is significant because it enables class members to practice participating in a community before engaging fully and directly in professional practice, before seeking to define their role in an external community. The goal for Wenger (1998) is to develop a shared sense of responsibility for the enterprise and its practice in order to learn about it and make it better.

Based on their understanding of *Terrarium* and the situatedness of the assignment scenarios, students emulate the development of a shared sense of responsibility by creating a detailed style sheet. To create a style sheet, students not only have to interpret the novel to understand the standards and expectations they must meet, they also have to continue the novel's story because it only provides so much narrative frame. As the work of the semester continues, students use the style sheet to guide their editorial actions. The standards themselves evolve as needed and as students become more situated and present within their consensual, shared reality. In other words, as their identities become more coherent, their understanding of the rhetorical situation, and their responsibilities within it, also become more coherent. That evolution is reflected in their work on the style sheet.

A community of practice approach works well in this context because it emphasizes participation as identity-building, and imagination as a mode of belonging. In the fifth edition of their *Technical Editing* textbook, Rude and Eaton (2011) mention imagination, emphasizing that editors offer more than "grammar and mechanics"; they offer "analysis, evaluation, imagination, and good judgment" (p. xix). This textbook provides, they say, a "process for imagining varied audiences" (p. xx). Although she does not elaborate, Rude (2010) suggests that "comprehensive editing requires imagination" (p. 54). Discussion about the role of imagination in editing, especially in terms of belonging to a community, is even more important in this age of CMSs.

## The Role of Stories

Wenger's (1998) discussion of stories and their role in a community of practice is vital for understanding his concept of imagination. Stories communicate the meaning of the practice in a COP. They provide what Miller (1979) identified as humanistic, the "concepts, values, traditions, and style which permit identification with the community and determine the success or failure of communication" (p. 617). In fact, the meaning of the practice is shared through the stories told by experienced members (those with full membership) as a process for sharing with new members (those with limited or novice membership) a picture of what constitutes competence in that practice (Wenger, 1998). This narrative picture of confidence guides inexperienced members in their progress toward creating an identity of participation. Inexperienced members tell their own stories in two ways. First, they reiterate the stories told by experienced members to communicate their understanding of the meaning of the stories. Second, they tell new stories to demonstrate new ways of looking at existing stories as a means for reinvigorating the practice and its value to the community with new perspectives. In this way, stories work as a method for defining the parameters and scope of a practice and its value to the community.

Because stories engage the imagination, members are able to "recognize [their] experience" in others and to create models that guide their ability to develop an identity of participation in the community, as demonstrated through their stories. It is, Wenger says, "imagination [that] creates a kind of community . . . to which [members] see themselves as belonging" (1998, pp. 181–182). The telling and retelling of stories keeps the imagination and its agency active. Members, both experienced and inexperienced, tell stories to demonstrate their ability to control a narrative, to frame what is and is not considered a plausible framework of competencies for the community. By working through the story and social setting of *Terrarium*, students learn to craft and control their own narrative frameworks (business and humanistic practices) for editing. The better able a member is to articulate the meaning of the practice through what I've called narrative ways of knowing, the better able the community works as a learning community (Bridgeford, 2004). My own story demonstrates the process and path I took to my understanding of the craft of editing, and the lessons I learned first through the imagination and then in practice.

## My Story

In 2007, when Bill Williamson, Karla Saari Kitalong, and I created *Programmatic Perspectives* for the CPTSC, I was strongly opposed to using a content management system to deliver the journal. At the time, I also served as the Information Officer for CPTSC, a role that involved maintaining the organization's website (CPTSC.org). I envisioned the journal's presentation to be embedded within, and thus reflective of, the design of that website, which I had re-created the year before. My traditionally crafted site replaced an existing site created in *Plone.*[1] As Information Officer, I was also charged with producing the annual proceedings for CPTSC. This project was one of the reasons the *Plone* site was built in the first place. The organization's executive committee at the time imagined a responsive environment that would facilitate the efficient compilation of the position statements into a single, final conference proceedings document. The system would have allowed committee members to review proposals, would have allowed users to revise proposals, and would eventually generate the proceedings, all in a fast, efficient manner.

This process is an activity well suited for a CMS such as *Plone.* But, as with the CPTSC journal and organizational website, I insisted on designing and producing the individual proceedings to reflect each year's particular theme. Like many authors unfamiliar with the way CMS-driven design works, I objected to using the *Plone* site because it would separate content generation from document design, and thus in my mind require me to relinquish control of the document (Clark, 2008). But I was wrong on several levels.

For one thing, a CMS certainly would have made my editing work for CPTSC more efficient, which would have been good for me as a busy academic. For another thing, I would have perhaps understood earlier the effect of CMSs on the field, both in industry and in academia. I might also have gained hands-on knowledge of CMSs, giving me more insight into work-related environments that would likely have, in turn, informed my pedagogy sooner. I also might have realized that thematic design was in fact well within my grasp if only I would have imagined my role a bit differently, and in so doing embraced the fuller capabilities of the system to which I had access. Finally, I was wrong because my imagination was hindered by the "craft" identity that has defined the field for so long, an identity that emphasizes "handcrafting unique document products one at a time" (Andersen, 2014, p. 119). In fact, many technical communicators in industry, Andersen (2014) argues, "still view themselves as independent crafts-people" (p. 121). Given this craft perspective, I couldn't imagine giving up design and document control at the time. Because I have always viewed my work from a craft perspective, that identity blocked my ability to develop a different world image of my work as an editor. Until I was able to imagine editing from a different perspective, I was not able to teach it by any means or framework other than as craft.

As an unintended consequence, I missed an opportunity to expand what it might mean for CPTSC (the journal and the organization) to function as a technologically collaborative community. From a communities-of-practice perspective, I effectively limited and centralized the organization's institutional memory, a significant era of which exists right now on my personal hard drive or in my mind. A community's institutional memory, what I have elsewhere called the narrative accrual (Bridgeford, 2004), contains the knowledge (domain) and social fabric (community) used to develop what Wenger (1998) calls an *identity of participation*. This identity is what members create so they might participate in the practices of a community competently. For editors, this identity traditionally involves communication with writers, copyeditors, proofreaders, and managers, all of whom are accustomed mostly to working in isolation, leaving red pen tracks along the way, so to speak. In my own experience as an editor of two journals and four edited collections, I learned how to interact mainly through email, primarily one-on-one, creating a community of only a few people at a time. In hindsight, if we (that is, the "we" of any of the projects I highlight here) had used a CMS, we might have drawn contributors and editors together to form a broader community-building framework, and, in doing so, we might have made the work of subsequent editors a lot easier. This experience shaped how I teach all aspects of technical communication, including editing.

## My Approach

Because stories can transmit knowledge quickly and easily, they work well to support classroom practices. The story in *Terrarium* works particularly well because its plot involves transforming society from a craft perspective to one of dynamic text. As a dystopian narrative, *Terrarium* is appropriate for any technical communication class because it involves new and different technologies that need to be defined and described for the world's citizens. For my classes where I use it, the novel acts as the students' (and thus as the community's) domain, providing information that members access through their imagination. The novel situates students in a rich environment, complete with all the relations of meaning and power that specific temporalities and social relations entail. All of this contributes to students' ability to engage in appropriate communication action and to build an identity of participation.

Narrative works well as a pedagogical agent because it is inherently interpretable (Bruner, 1991). That is, it can't be read without also being interpreted. *Terrarium* offers a plausible fiction in that it frames a context and exigence that represents a realistic possible future. I have used it in all my courses at one time or another. It is a quick read with characters who are easy for students to identify with. Briefly, *Terrarium*'s plot focuses on a heavily polluted Earth that has become inhospitable. To survive, the remaining people build huge, globe-like enclosures to safely house humanity. The society in *Terrarium* grapples with new and different technologies that need to be defined and described. I set my writing and editing students' tasks related to this mission.

Because imagination "requires an opening" (Wenger, 1998, p. 185), I use *Terrarium* to present a plausible community and to help students frame a basic understanding of the world within, which I ask them to operate for class projects. The novel transmits relevant knowledge, and "can be appropriated easily" by students, thus allowing them "to enter the events, the characters, and their plights" (p. 203). Wenger notes that stories "can transport our experience into the situation they relate and involve use in producing the meanings of those events as if we are participants" (p. 203). Thus, they help students to see their own and others "engagement through the eyes of an outsider" (p. 185), a task Wenger says is essential for building what he calls an identity of participation. *Terrarium* is a safe environment for them to practice the skills they need but also to imagine themselves as being an editor. Here, students have permission to explore and discover, to succeed and fail, in ways that help them keep one foot in the story and one foot in reality while trying on the role of editor.

Because I have used *Terrarium* in classes since 1998, I have compiled a large archive of student documents (e.g., reports, memos, instructions, process descriptions, definitions, and proposals). This repository offers another kind of CMS that I share with students. I use this archive in conjunction with the novel itself to create assignment scenarios. For example, the following broad scenario serves as a place for students to start thinking about themselves as editors.

> **Scenario.** It's the year before Enclosure Day (2025) and the Institute of Global Design (IGD) is planning for the movement of humanity into the enclosures. Given that all daily activities will require learning new technologies and processes (e.g., using a vidphone, operating a vaporizer, submitting a health report, etc.), Zuni Franklin, lead architect at IGD, has been charged with adapting existing print content into dynamic text for a central content management system, Cybernet. Cybernet will house all content for the appliances, machinery, and processes new to the enclosure and will be accessible from all terminals throughout the enclosure. As part of Zuni Franklin's editing team, you will be editing and preparing these documents for display in various devices for the Cybernet.

Like all scenarios, this one works by providing students with a human problem—moving into the enclosure. Some of the documents I use from previous classes were constructed originally as whole documents (e.g., user manuals). Students edit those documents, crafting them into topic-based chunks of content. This work draws on their imaginations first by situating them within the novel's story, and then by recreating the document from a dynamic perspective. Because the project is collaborative, it situates students not only as editors within the novel's story (and their own), but also as traditional editors moving from a craft perspective to a dynamic content perspective. I typically will choose an existing character or character for whom students direct their editing. For example, in the scenario above, I situate students as part of the editorial team for Dr. Zuni Franklin, a major character who takes a leadership role in the novel. Students must imagine the primary audience as residents of Oregon City (one of the enclosure cities in the novel) and one of the secondary audiences as Dr. Franklin, the supervisor. Although the audiences are imagined, the demands of editing for consistency, continuity, accuracy, and grammatical correctness are all very real. This approach keeps the focus on people, what Pullman and Gu (2009) call a technologically critical approach that views content management as "people managing content/information" (p. 4). Because "a CMS is ultimately not about software; it's communication" (p. 7). The value of this approach is the focus on organizational and communicative contexts.

## Imagination as Agency

In communities of practice, creating images, the "work of imagination" (p. 177), is characterized by Wenger (1998) in nine ways. I point out here the two most relevant for this chapter. Wenger says that part of belonging through imagination involves sharing stories, providing explanation, or describing situations. He emphasizes that members do this by generating scenarios that help other members explore other ways of doing, other possible words, and other identities (see p. 185 for the entire list).

Thus imagination works through stories because they function as an "important source of identification" (Wenger, 1998, p. 194). That is, they help shape images of the practice that empower members to participate and contribute, and thus foster an identity of participation. Because stories are one of the ways we appropriate meaning, identity-building occurs in this context first through the hearing and telling of stories (Wenger, 1998, p. 194). Those stories enable community members to build confidence both when listening to other members' stories as well as when telling their own stories. The meaning of the practices defined within the narratives shared by experienced members (those with full membership) is thus passed on to inexperienced members (those with limited or new membership), who gain a picture of what constitutes competence in that practice. This narrative instantiation of confidence guides inexperienced members in their progress toward creating their own identity of participation, which is Wenger's (1998) notion of agency. In this way, stories work as a method for defining the scope of a practice and its value to the community. Thus, imagination enables members to "recognize [their] experience" in others, and to "create models" that guide their ability to develop full membership in the community. It is, Wenger says, "imagination [that] creates a kind of community . . . to which [members] see themselves as belonging" (pp. 181–182). The telling and retelling of stories keeps the imagination and its agency active.

Imagination's work in my classroom is not merely an act of fantasy, although that contributes much to students' abilities to situate themselves realistically in the scenario. The scenario is more about providing source material to help students develop a sense of belonging. Just as they would for any real-world company, upon the very beginning, students would have to imagine themselves as editors. Imagination's role in belonging has always been a process of enculturation. If students are to be editors, they must see themselves as editors. Students have to imagine that they belong, that they are comfortable in the editorial role, comfortable with the notion of occupying such a position of authority. In the beginning, students have a tough time imagining themselves as the kind of professional to whom another person should listen. This aspect of teaching students to be editors goes beyond much of what they expect to learn—editorial tasks—and encourages them to see themselves within a context in which they are the experts. Once they have an identity of participation, they will know they have gained status as a full member of a community. This status aids them in affecting change to an author's text (or querying the author about a change). Changing someone's text takes an act of courage that imagination can empower them to act. In this way, students have to make sure that the editing practice—what it means—is sustained as part of a community's values. This whole concept of confidence, then, begins with imagination.

In an editing class, students need to learn not only the basic duties of an editor—copyediting, global revision of content, proofreading, and so on. They also need to know how to become editors as organizational entities, how to develop into that authoritative position, and how to identify with and belong to the distributed community accessing the CMS. To recast Miller's (1979) framing of technical writing for editing: "to [edit], to engage in communication, is to participate" in the practices of a community, and "to [edit] well is to understand the conditions of one's own participation" (p. 617). We can teach students the mechanical fundamentals of editing, but we must also provide some kind of broader social, cultural context (other than the classroom) to teach them how to see the editor role as a form of community agency, and thus how to belong to a community. Wenger's concept of an identity of participation can help us help students develop as strong editors. To develop this identity, students must be "able to make distinctions between reified standards and competent engagement in practices," which is "an important aspect of becoming an experienced member" of a community (Wenger, 1998, p. 82).

For example, with *Programmatic Perspectives* (*PP*), Bill, Karla, and I agreed that unless an article was distinctly not written for *PP*'s audience, we would not outright reject any manuscript. This practice grew out of our belief in mentoring writers as part of our editorial responsibilities (Kitalong, 2009). Recognizing why we do that is the difference between serving as gatekeepers for the scholarly community or as agents of enculturation. Such philosophical stances shape intellectual communities. Building an identity of participation happens when members can identify their and others' ability to take "ownership of the meanings" they create, that is, they understand how to contribute to the enterprise of the community in ways that add value to its meaning.

Students have succeeded when they put forth a "claim to owning the meaning of a piece of text, a knowing smile, a tool, or an idea is being able to come up with a recognizably competent interpretation of it" (Wenger, 1998, p. 201). By text or tool, Wenger does not mean members take possession of a specific artifact. Rather, he means ownership of a meaning that "has currency" to the community's enterprise and that is recognized as a "legitimate contender" (p. 201). The meaning created becomes part of how members look at and interpret the enterprise. At this point, members gain full membership status, that is, they understand who they are in this community, who others are, and especially, the value of their own and others' contributions. They "understand," then, "the conditions of [their] participation," as Miller (1979) said. When this happens, the humanistic mission has been fulfilled.

To stay focused on the humanistic aspects of editing, the course needs to have a context that enables agency to develop. In this time of CMSs, agency can seem nonexistent because the technological framework seems daunting, decentralizing, and even dehumanizing. It becomes tempting to focus more on the technology and the process of editing than on the broader context of social and professional roles, of real agency. If all students learn to do is manipulate sentences, they never reach beyond the mechanistic. They need broader thinking skills about how meaning is built, and how a document contributes to the content in the database, and how the database contributes meaningful knowledge to the community. Wenger's (1998) notion of imagination as agency helps us cast ourselves forward into alternate conceptualizations and contextualizations of the design process. Wenger grants us both the permission and the pathway to expand our expertise. Enculturation is the permission to see what is; imagination is the pathway to see what can be.

## Editing and Content Management Systems

A responsive and relevant editing pedagogy must emphasize knowledge and skills for working with content management systems (CMSs). To assert that CMSs have changed how we work in the field of technical communication is an understatement. Although the traditional tasks for editing have not changed, in that editors still engage in activities such as comprehensive editing and copyediting, the processes by which that work gets done have changed. Pullman and Gu (2009) describe a CMS as consisting of "a database containing data and metadata and a Web template that controls the 'look and feel' of information" (p. 2). A CMS's technological frame can seem intimidating to teachers who draw their expertise from the craft tradition in that it seems to remove the human element (Lanier, 2012). Editing in a content management system involves editing distinct chunks of information that have been written as topic-based content. The writing and editing of documents (or topics) may be completed directly in the environment of the CMS. Dynamic content, that is, content stored in a database, is drafted and tagged using extensible markup language (XML), a coding schema that identifies the kind of content being stored as well as when and where it will be placed when called for by a user. The system that develops within the context of the CMS enables editors to develop an agency network, that is, a system of actions culled from collaborative editing practices. The CMS software then assembles the content when requested by a user according to the output parameters contained in the code. The topic-based chunks of content pulled together by the software need to make sense both on their own and as part of a complete document. I use *document* loosely here because how content is rendered once assembled by a user depends on the device on which the access occurs, as well as on the search terms used. To work in this technological framework requires the work of imagination because there may never be a document in the traditional sense.

In a CMS, the work of imagination is active in two ways. First, imagination plays a significant role in editing topic-based content because editors need to know how the single topic fits within any complete document, regardless of how such documents might be rendered and published. When editing a topic, editors need to see not only the immediate text, but also how it might become a chunk of content to be displayed on multiple platforms, thus enabling its reuse across multiple media platforms. This second element is not immediately visible in a CMS, because individual topics are tagged and stored separately. Second, editors must be able to imagine readers using the documents. In these two ways, imagination plays an active role in how editors visualize a document in a CMS, how they act on that document, and how users use the document in context. Because the concept of *document* is more ephemeral in a CMS, imagination is a constant activity for editors, who must continually visualize the complete content set, and anticipate the users' needs and expectations of that content.

Stories such as *Terrarium* help us ensure that humanistic values are central to transmitting knowledge.[2] Stories are an "important source of identification" (Wenger, 1998, p. 194), helping shape practice, and thus empowering members to participate and contribute, and fostering an identity of participation. Because stories are a means of appropriating meaning, identity-building occurs through the hearing and telling of stories. Wenger places great value on stories as a means for communicating meaning in the community, which enables the building of confidence both when listening to stories as well as when telling stories within the community.

The document database I provide for students situates them within the context of the craft perspective, because these documents were written and designed from the more traditional view where "both paper and online documents exist as a single document which an editor reviews before it reaches the reader" (Albers, 2000, p. 194). This context means that the documents were created in an owner-specific manner that must be revamped to a dynamic content system. This use parallels the craft versus dynamic shift ongoing in industry today (Hackos, 2000). The topic-based content that students edit puts them within a hypertextual or dynamic context perception, requiring them to take craft documents and recraft and edit them for a dynamic system. For example, I used to ask students to edit an entire document, but now I have them work with topics, or chunks, of content that must both stand alone and be part of a "whole" document. I used to put students into groups and ask each group to edit an instruction set that had previously been edited as a coherent whole by another group. I now ask groups to identify topics within that instruction set, and to edit accordingly.

The work evolves in stages in my editing classrooms. Before students revise any texts, we first determine as a class the best writing structure for the topics. They then distill the standards for those structures into a single style sheet. They work together in Google Docs to identify and edit the various topics within that manual. Each group edits a topic for a different medium. They then physically relocate to the computer another group used for its work, and, using the style sheet, assess their peers' work. Because at my institution I don't yet have a working CMS, I emulate the dynamic nature of the editing tasks with physical movement to get them to see that content is everywhere. The craft model of composition teaches us that sharing content is plagiarism, and thus poor practice. The dynamic model of composition instead encourages us to think about sharing content as effective collaboration. Being able to differentiate the two can help students adjust to the fluid nature of content in a CMS. This role, at one time more craft-minded, has become more dynamic, and not only a technological change for the community, but a narrative change as well. Imagination affects the narrative framework students build to help them see the way to building an identity.

An editor's job has always been to decide when, where, and how to focus on and produce a final document. When do editors focus on little pockets of content, and when do they look at the complete document? Because topic-based content is written as stand-alone chunks of text that work both independently and cooperatively in the technological and human systems, editing students must learn how to understand the larger whole (whether that means a document or an interface), and to decide when that view is appropriate. If students remain fixed on single-author ownership of complete, linear documents, they will be too invested in the craft perspective. If so invested, they will lose sight of the larger professional and cultural transformations that have become so significant to the work of technical writing and editing outside the academy. As well, if they become too focused on design, they will lose sight of the content. When students can transition to designing and contributing to dynamic information systems and see text as distributed throughout an organization and across multiple projects, they might begin to assume the role of editors in a CMS-driven environment. Eventually, they have to imagine complete documents, or at least the range of possible configurations for a database. Tagging content as they go along can help students learn to envision the entire document while editing an individual topic. The content audit, which is "an accounting of the information in [an] organization" (Rockley & Cooper, 2012, p. 102), is also useful in this respect. The purpose, Rockley and Cooper describe, is to "analyze how content is written, organized, used, reused, and delivered to its various audiences" (p. 102). Having students conduct a content audit helps provide them with a picture of the entirety of the organization's documents, document structures, and topics.

It is scary for students to transform the way they think from craft to responsive composition models. Part of what makes it scary is that they are used to controlling the content. But when they must work with responsive, shared texts, that control is distributed throughout the CMS, and throughout the organization. It is important for students to learn to relinquish the control (and with it the sense of individual authorship) they're used to with craft documents.

Giving students a narrative context—a framework—helps them learn to relinquish control in positive ways. Good science fiction works particularly well because it must be plausible to engage readers. Students might readily identify with the story and the characters. If the narrative an instructor chooses is inauthentic, students won't buy into it. Likewise with users of CMSs; if the story the content tells is not plausible, users will simply click away. The narrative brings with it a plausible set of possibilities. Students can see themselves belonging to a community in which they are the experts and for reasons within and beyond copyediting and proofreading. They learn to belong and therefore learn how to construct an identity of participation. Regardless of the platform, members of a community need to understand themselves as belonging, and our pedagogy must remain committed to helping them earn that understanding. But the craft tradition cannot and need not be left behind. We need to expand our conception of it, because craft-tradition linear documents remain part of our technical communication imagination. Recognizing the existence and value of dynamic text is merely part of an evolutionary stage. In doing so, we recognize that responsive documents exist on a spectrum that includes at one end traditional print documents, and at the other an array of dynamic, interactive genres for communication. These documents may not be linear, but they are not truly incomplete, as they might at first seem to uninitiated designers (or teachers). Perhaps people resist—just as I resisted in the beginning—because these genres feel disconnected, amorphous, and yet at the same time manufactured. But it's important to remember that all text is manufactured.

Driven by this pedagogical framework, we can ensure that the work of imagination is central to teaching students how to be editors. Imagination is important to pedagogy because it empowers students to act as they learn how competence is defined in a community of practice. We want students to walk out of our classes with the confidence not only to be editors but to be editors who can work in a CMS environment. Whether teachers have a fully functioning CMS to use in their classroom or are simulating a CMS through services such as Google Drive, they can effectively create an atmosphere that will introduce students to editing as part of a community. What's at stake is students' ability to adapt to the CMS environment, as Pullman and Gu (2008) argue: "We cannot in good conscience graduate people in technical communication who are not prepared to work in these environments" (p. 4). Driven by this pedagogical framework, we can rethink how editing gets taught in an individual course or within the context of other courses. What's needed are pedagogical approaches that describe in some detail the ways imagination can work in writing, editing, and design classrooms. We also need more research that shows actual editing in industry contexts such as Kreth and Bowen's (2017) descriptive survey of technical editors. Finally, we should find ways to collaborate with industry partners so we can learn from and inform each other's work. The work of imagination is vital to our editing future.

## Pedagogical Applications

Structurally, this pedagogy is woven into the communities-of-practice framework. In this section, I focus only on the top level of analytical processes, including situating action through narrative ways of knowing, creating community through imagination as a mode of belonging, and establishing acting positions through scenarios that invite students to enter. In this way, the pedagogy enables imaginative agency through activities associated with a CMS such as refashioning documents into topic-based content. The application of this pedagogy addresses the novice teacher in constructing a framework for editing activities in ways that produce new images of the community in action. Given that the regular aspects of teaching editing (copyediting, etc.) are covered fully elsewhere (Albers, 2000; Rude, 2006), I focus here on the larger pedagogical narrative-ways-of-knowing approach described in this chapter.

### Establishing Narrative Ways of Knowing

To establish a narrative environment that fosters imagination as agency, you need to first create a context within which students will be working and completing assignments, as well as using to begin imagining themselves as editors. Whatever narrative context you choose, it should be something that can be read quickly and understood easily. Keep in mind that students will not be reading the narrative as a literary artifact; rather, it's used as a context for assignments. I chose *Terrarium* because it is a quick read and because it is easy to apply technical communication principles and practices to the world it imagines. Even though we are not reading it from a perspective of literary traditions, we do discuss as a class the social and cultural meaning and significance of events in the novel so that we come to a shared understanding of the rhetorical situation within which we work. It's our first effort to come together as a community and develop our editing practices through narrative ways of knowing.

For example, through the vignettes in *Terrarium*, we learn that Enclosure Day takes place in 2026, thus marking the moment an act of legislation required everyone to move into the Enclosures. I cast students as editors for the new Enclosure Board. Students are tasked with the responsibility of recreating complete print documents developed prior to life in the Enclosures as content suitable for management through a CMS. To do this, they must first review the content for the Enclosure Cybernet system that will be accessed by "vidphone" (smart phone), "vidscreen" (TV screen), "vidplay" (tablet), or "vidstation" (kiosk) (as they are described or imagined in the novel). Students begin creating personas to represent the audiences who need the information. Mostly, I don't allow students to use existing characters from the novel for two reasons: (1) students are tempted to stay within the confines of the novel, and (2) invention is part of creating an identity of participation. By imagining personas new to the story, students simultaneously engage in a narrative way of knowing and create their own identity of participation.

## Conducting a Content Audit and Creating a Style Sheet

Following audience analysis, I ask students to conduct a content audit from my collected repository of documents. An audit, or an "accounting of an organization's collection of documents" (Rockley & Cooper, 2012), identifies the viability of content in conjunction with the organization's needs as well as identifying topics and their granularity. In doing so, students look for "opportunities for reuse" (Rockley & Cooper, 2012, p. 111). Through the assignment scenario, I describe the exigency of a situation and categorize what students are being asked to do. This scenario provides a typical example from my class:

> It's 2025, the year before Enclosure Day, and humanity is in the process of preparing to move into the Enclosure. The Enclosure Board has asked your group of editors to identify the viability of existing content before its move to the Enclosure Cyberboard, a centralized content management system. Access the document repository for your group's assigned content.

The repository resides in Canvas or Google Drive, demonstrating the specifics of a content audit in class first by demonstrating it and then by putting students into groups. Each group works with a separate subset of randomly chosen documents. In choosing documents for students to edit, I first remove all identifying information from the files (including any summary information that may have been retained in Microsoft Word), and then provide the content written by previous students in varying stages of competency. Based on Rockley and Cooper's (2012) approach to top-level analysis, I ask students to create a table in which they identify information products by media (see Table 4.1). They align their audit spreadsheet using language that works with *Terrarium*. Thus, the results of a content audit for vaporizers might look something like Table 4.1, which shows the situatedness of the content required for each medium.

Students must assess the quality of the content, determine the appropriate level of edit required, and decide what it will take to transform that content into topics. Once students have identified the topics and edited them as they would for any craft document, I ask them to identify topics and create a style sheet for editing topic-based content. It is important at this point for them to differentiate between refining the text as stand-alone content, and editing the content for granularity, or reuse. As a class, we create and refine this style sheet to identify the structure to follow when editing topic-based content. For example, Figure 4.1 shows a section of a style sheet based on *Terrarium* and the scenario I created identifying the hierarchy and section names for the XML topics for a topic, "Submitting a health report" (mentioned in the novel). In *Terrarium*, residents are required to submit a health report if they witness a violation, so this kind of chart makes sense to include as part of a style guide.

Creating this topic hierarchy enables students to imagine the content and the action it depicts. Students must invent the language used for the XML topics, which helps them imagine the tree and its various branches of the hierarchy. The branch begins with the type of report and the type of violation being reported. From that branch, an editor would identify and organize topics based on their status as major or minor violations. Each major or minor violation is then broken down into content chunks: the offense that occurred, a description of the incident, and if it is a first or repeated offense. The structure of this

**TABLE 4.1** Top-level content audit results for a vaporizer

| Content | Vidscreen | Vidtablet | Vidphone | VidKiosk |
|---|---|---|---|---|
| Enclosure logo | X | X | X | X |
| Contact information | X | X | X | X |
| Vaporizor definition | X | X | X* | X |
| Vaporizor description | X | X* | | X |
| Setting up the vaporizer | X | X | X | |
| Inserting trash | X | X | X | |
| Caring for your vaporizer | X | X | | |
| Solving problems | X | X | X | |

*The same content is used but in a shorter version.

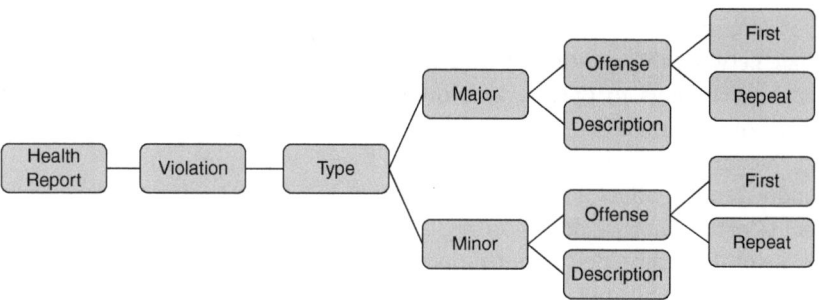

**FIGURE 4.1** XML topics for submitting a health report.

reporting tree shows the relationships and their connections among the various parts of the description. In a style sheet, identifying this hierarchy is important for editors in instructing technical writers on the organization of the topics. Creating this hierarchy takes a significant amount of imagination and the narrative helps students enter that space to do so.

With these kinds of assignments, students are not only using their imaginations to see content differently, they are also using content to help see themselves as editors, to create an identity of participation that helps them develop the authority needed to impose change on someone else's prose. Entering the narrative context to edit can help them recognize the role of the editor and imagine the possibilities available to them. Once they can imagine, they can be editors.

## Conclusion

By using narrative ways of knowing, students can safely explore what it means to take on an editor's role in a community of practice. Although students have at times resisted reading a novel in a technical editing course at the beginning of the semester, this practice is consistently praised in students' course evaluations, because the context helps them imagine more completely the needs of the audience. Of course, instructors might choose instead a short story or another kind of narrative text. The medium or genre is not what is important. The goal is to help students develop an identity of participation through editorial acts while honing the skills they need to edit documents in a content management system. Any plausible, accessible narrative might thus serve well. The ease with which students might immerse themselves in a new consensual reality is key. Transforming traditionally crafted, coherent documents into dynamic, topic-based source content helps students shift from craft to dynamic content models of composition and design. Such a course frame serves for instructors as a transitional pedagogy that might help to expand our notions of both craft and dynamic texts and pedagogies.

Using communities of practice as a writing strategy creates a transitional state between the tradition of craft and the use of dynamic content. This space enables students to operate in economies of practice, thus reimagining the value of content, composition, editing, design, and, ultimately, of how we create meaning through text. Using these back-and-forth approaches for discovery and invention offers a stepping stone for creating agency-driven pedagogy. It is important that we don't lose sight of what the craft tradition values in the wake of the technological innovation offered by content management systems. Instead, we need to bring together craft and dynamic models of composition to reimagine and revalue composition as a whole. Although editors will always engage in close refinement of texts, they must learn to do so within a dynamic system that privileges new ways of using (and reusing) content. The approach I describe in this chapter is one configuration that works toward that change.

## Pedagogical Practicalities

- Find a suitable narrative context for the documents you will ask students to edit and develop assignment scenarios from them. In addition to *Terrarium*, I have also used *Dune* and *The Hunger Games*. An instructor I know uses *The Time Machine* in his technical writing classes.
- Accumulate a document repository for content audits. Although few people may begin with a decade's worth of student writing as I did, you could repurpose any documents from previous classes.
- Practice writing topic-based content to develop your competency.
- Establish a framework for a content management system. If your university does not offer this capability directly, consider using a wiki, Google Docs, or WordPress.

## Acknowledgements

I wish to acknowledge the superb feedback I received from my colleagues Bill Williamson, David Peterson, and Kathy Radosta.

## Notes

1 *Plone* is an open source software for creating content management systems. The *Plone* site was originally created by Jim Ridolfo who was at the Writing, Information, and Digital Experience (WIDE) center at Michigan State University at the time.
2 For example, Julian Orr's (1996) research on the practices of Xerox repairpersons demonstrates the same value placed on storytelling in a community. The repairpersons met weekly to discuss what they learned about repairing Xerox copiers, because answers to copier problems were not always available in the technical manuals. They framed their knowledge in the form of stories, communicating both what to do and what not to do as well as how to be and how not to be a member of the community.

# References

Albers, M. (2000). The technical editor and document databases: What the future may hold. *Technical Communication Quarterly, 9*(2), 191–206. doi:10.1080/10572250009364693

Andersen, R. (2014). Rhetorical work in the age of content management: Implications for the field of technical communication. *Journal of Business and Technical Communication, 28*(2), 115–157. doi:10.1177/1050651913513904

Bacha, J. (2009). Single sourcing and the return to positivism: The threat of plain style, arhetorical technical communication practices. In G. Pullman & B. Gu (Eds), *Content management: Bridging the gap between theory and practice* (pp. 143–160). Amityville, NY: Baywood.

Baker, M. (2013). *Every page is page one*. Laguna Hills, CA: XML Press.

Blyler, N., & Thralls, C. (1992). *Professional communication: The social perspective.* Thousand Oaks, CA: Sage Publications.

Bridgeford, T. (2004). Story time: Teaching technical communication as a narrative way of knowing. In T. Bridgeford, K. S. Kitalong, & D. Selfe (Eds), *Innovative approaches to teaching technical communication* (pp. 111–134). Logan, UT: Utah State University Press.

Bridgeford, T. (2007). Communities of practice: The shop floor of human capital. In C. Selfe (Ed.), *Resources in technical communication: Outcomes and approaches* (pp. 161–178). Amityville, NY: Baywood.

Bruner, J. (1990). *Acts of meaning.* Cambridge, MA: Harvard University Press.

Bruner, J. (1991). The narrative construction of reality. *Critical Inquiry, 18*(Autumn), 1–21. doi:10.1086/448619

Clark, D. (2008). Content management and the separation of presentation and content. *Technical Communication Quarterly, 17*(1), 35–60. doi:10.1080/10572250701588624

Dayton, D. (2003). Electronic editing in technical communication: A survey of practices and attitudes. *Technical Communication, 50*(2), 192–205.

Hackos, J. (2000). Trends for 2000: Moving beyond the cottage. *Intercom, 47*(1), 6–10.

Johnson-Eilola, J. (1996). Relocating the value of work: Technical communication in a Post-Industrial Age. *Technical Communication Quarterly, 5*(3), 245–270. doi:10.1207/s15427625tcq0503_1

Johnson, R. R. (1998). *User-centered technology: A rhetorical theory for computers and other mundane artifacts.* New York, NY: State University of New York Press.

Kitalong, K. S. (2009). Mutual mentoring: An editorial philosophy for a new scholarly journal. *Programmatic Perspectives, 1*(2), 211–216.

Kreth, M., & Bowen, E. (2017). A descriptive survey of technical editors. *IEEE Transactions on Professional Communication, 60*(3), 238–255. doi:10.1109/TPC.2017.2702039

Lanier, C. (2012). Accounting for the human element when planning for a content management system. *Technical Communication, 59*(2), 99–111.

Longo, B. (2000). *Spurious coin: A history of science, management, and technical writing.* New York, NY: State University of New York Press.

Miller, C. (1979). A humanistic rationale. *College English, 40*(6), 610–617. doi:10.2307/375964

Moore, P. (1996). Instrumental discourse is as humanistic as rhetoric. *Journal of Business and Technical Communication, 10*(1), 100–118. doi:10.1177/1050651996010001005

Moore, P. (1999). Myths about instrumental discourse: A response to Robert R. Johnson. *Technical Communication Quarterly, 8*(2), 210–226. doi:10.1080/10572259909364661

Orr, J. E. (1996). *Talking about machines.* New York, NY: ILR Press.

Pullman, G., & Gu, B. (2008). Guest editors' introduction: Rationalizing and rhetoricizing content management. *Technical Communication Quarterly, 17*(1), 1–9. doi:10.10 80/10572250701588558

Pullman, G., & Gu, B. (2009). *Content management: Bridging the gap between theory and practice.* Amityville, NY: Baywood.

Rockley, A. (2003). Single sourcing: It's about people, not just technology. *Technical Communication, 50*(3), 350–354.

Rockley, A., & Cooper, C. (2012). *Managing enterprise content: A unified content strategy* (2nd ed.). Indianapolis, IN: New Riders.

Rude, C. (2006). *Technical editing.* New York, NY: Pearson.

Rude, C. (2010). The teaching of technical editing. In A. J. Murphy (Ed.), *New perspectives on technical editing* (pp. 51–65). New York, NY: Routledge.

Rude, C., & Eaton, A. (2011). *Technical editing* (5th ed.). Allyn & Bacon Series in Technical Communication. Boston, MA: Longman.

Sanders, S. R. (1985). *Terrarium.* Indianapolis, IN: Indiana University Press.

Smith, E. (1997). Intertextual connections to a "Humanistic rationale for technical writing." *Journal of Business and Technical Communication, 11*(2), 192–222. doi:10.1177/1050651997011002003

Sullivan, D. (1990). Political-ethical implications of defining technical communication as praxis. *Journal of Advanced Composition, 10*(2), 375–386.

Tebeaux, E. (1980). Let's not ruin technical writing, too: A comment on the essays of Carolyn Miller and Elizabeth Harris. *College English, 41*(7), 822–825. doi:10.2307/376223

Thralls, C., & Blyler, N. (1993). The social perspective and pedagogy in technical communication. *Technical Communication Quarterly, 2*(3), 249–270. doi:10.1080/10572259309364540

Wenger, E. (1998). *Communities of practice: Learning, meaning, and identity.* New York, NY: Cambridge University Press.

Wenger, E. (2015). Introduction to communities of practice: A brief overview of the concept and its uses. Retrieved from http://wenger-trayner.com/introduction-to-communities-of-practice

# 5

# TEACHING EDITING THROUGH A FEMINIST THEORETICAL LENS

*Susan L. Popham*

## Chapter Takeaways

- The practice of editing is always infused with theory, generally that of rhetorical theory, even if the theoretical foundation is unacknowledged.
- In tandem with rhetorical theory, feminist theory is well suited to support the work of editing and the teaching of editing because feminist theory encourages

  o a more conscious attitude of respect for authors and their manuscripts;
  o a more conscious analysis of female-based metaphors often used in describing editorial work; and
  o a more realistic exploration of the placement and hire-ability of women in editing and publishing workplaces.

- Feminist theory when taught as a theoretical foundation in the editing classroom may help editing students acknowledge other theoretical principles underlying the work of editing.

Much like the traditional view of technical communication as objective, clear, and plain (Buehler, 2003), technical editing is often viewed as an objective, neutral, pragmatic work, devoid of pesky or troubling theoretical approaches. In fact, only a few editing scholars (Buehler, 2003; Dragga & Gong, 1989) directly explore theoretical approaches to editing. However, despite such a lack of theoretical perspectives in editing scholarship, it is clear, when exploring the history of editing practices, that some theory does affect the work lives of editors.

# Editing Theory

In his history of technical editing, Warren (2010) notes that throughout the last four decades of the twentieth century, various scholars had cast their scholarship in tones of value judgments in editorial work: Lytle, in 1960, emphasizes the "feel" that an editor has for the reader (p. 33) that is usually based on the editor's own experience; such a value judgment is synonymous with a theory of intuition and affinity born of personal experience. In later decades, according to Warren, other editing scholars and teachers began to promote and call for an increased attention to the audience of edited texts. Such a recognition of audience is directly connected to an awareness of the importance of audience in rhetorical theory. Despite Warren's acknowledgement of these scholars' arguments, neither the scholars he cited nor did he recognize that such arguments were in fact based in an assumed theoretical foundation. In other words, they argued for the importance of recognizing specific practices, but none of them, including Warren, directly called such practices theoretically based. Perhaps one of the reasons why theoretical approaches to editing are rarely explored is that rhetorical theory pervades the entire editing enterprise; rhetoric, that is the study and application of textual and purposeful strategies for persuading audiences through textual (and contemporarily, extratextual) means is at the heart of any editing goal—to help the author effectively persuade one's audience through the best textual means. In other words, the art of rhetoric is the theoretical essence of editing.

One of the fundamental texts in editing "Situational Editing: A Rhetorical Approach" by Mary Fran Buehler (1980/2003) adeptly ties theory to practice in her description of technical editing:

> Many of us, I believe, have thought that the clear, plain, objective style that we normally strive to achieve is not rhetorical . . . But the truth is that the spare, objective style of technical writing is, in itself, a rhetorical choice.
> (pp. 459–460)

Her theory, originally published in 1980, is of such value that it was chosen for reprinting in 2003. She argues that a rhetorical approach, specifically that which works through a situational lens, is most necessary for the editor. She claims that while knowledge of grammar and other mechanics of discourse, like spelling and punctuation, what she calls "programmatic approach," is necessary for all editors, such knowledge is not enough. Rather, she argues, editors need to address language in the varied situations that might arise; a rhetorical theoretical approach more appropriately addresses situational challenges. Buehler concludes:

> The editor's position in technical communication is unique since the editor functions in the center of a series of rhetorical situations, linking the author and the potential reader, and serving the needs of both. The editor

cannot solve the problems of effective communication by using programmatic techniques alone. The rhetorical approach . . . offers greater rewards in terms of effective communication.

(p. 463)

Shortly after her original publication, another fundamental text, *Editing: The Design of Rhetoric* (Dragga & Gong, 1989), made clear the point that editing is necessarily rooted in rhetoric:

Rhetorical theory provides the basis or philosophical foundation for the editor's judgments. . . . We believe theory must inform practice. Without this correspondence, the editor is merely a technician, given to the blind adoption of rhetorical guidelines rather than their insightful adaptation.

(pp. 11–15)

In their pedagogical treatise on editing, Dragga and Gong go on to explain how the rhetorical canons of invention, arrangement, style, and delivery can help to elucidate and inform the work of the editor as he or she helps the author effectively shape a text to satisfy the desires of the audience. They point out that, as much as "audience" is a crucial editorial and rhetorical tenet, it too is a term in need of a greater nuanced understanding; simply invoking "audience" like a god-term in editing and in rhetorical practice is unlikely to achieve a suitable correspondence of theory informing practice. Instead of simple invocations to remember the audience, they argue that editors and editors-in-training must understand how to address a "real audience," rather than relying on a fictionalized, abstract view of a distant readership.

These two early works in the scholarship of technical editing are admirable in making clear the entanglements of, and disentangling, rhetoric with editorial practice and the need to recognize the theoretical bases of our practices. As editors and teachers, we need to explore more theoretical underpinnings of our editing practices and teaching practices—not just rhetorical theory, which was so aptly done by Buehler (1980/2003), and Dragga and Gong (1989)—but also to explore other theories as they connect to the work of editing. As an example, Ball (2018) explores how other theories, like genre theory, multimodal theory, sociocultural theory, and design theory help to explain her editorial practices and philosophy. On her personal website, she describes her conglomeration of theoretical approaches:

Applying multimodal and rhetorical genre theories . . . also means drawing on my experience teaching and researching workplace writing, web design, print design, and information studies in technical and professional communication; aesthetics, poetics, and hypertext theories in literature and creative writing (particularly as those intersect in electronic litera-ture); histories of print- and screen-based text production and delivery in media studies and textual studies; digital media practices in art and design; and other disciplinary areas and research. All available means. This con-glomeration is where I find myself: focusing inwards, toward a "home" field of digital writing studies while simultaneously focusing outwards, toward digital publishing studies as a specialty that embraces the collabo-rative, open-access, and professional values of the digital humanities.

Her philosophy goes far towards advancing an argument that editorial prac-tice is essentially entwined and founded in multiple theories, most notably that of situational rhetoric. However, despite these approaches to editing by Buehler (1980/2003), Dragga and Gong (1989), and Ball (2018), little scholarship has since been published that directly examines and argues for specific rhetorical and/or theoretical approaches in editing practices and editing classrooms.

I begin by laying bare a basic assumption, one that I hold strongly and will use throughout this chapter: All practice is informed by theory. Most scholars would agree with me, and in fact several scholars, like Dragga and Gong (1989), made that very argument long before me. That practice is informed by theory is considered a truism of contemporary scholarship, and thus it needs no bolster-ing from me. I only make this claim clearly here so that readers will not have to dig far in order to understand the assumptions that undergird this work.

However, like Ball, we must go further than simply acknowledging rhetorical theory. We must provide specific methods for teaching theory, and we must acknowledge other theories highly relevant to the work of editors and editing students; for me, feminism. Here is another basic assumption: Feminism—that is, writ broadly, a belief in the equal and respectful treatment, pay, and recogni-tion of women, while sometimes also calling attention to venues in which such respect was and is denied women—is an important, essential theory to bring to all classrooms, including the editing classroom. It is not the intent of this chapter to explore all the important aspects of feminist theory, like third-wave feminist theory, psychological feminism, or decolonial theory, an exploration that would likely take many books; however, this chapter's view of feminism does align most closely with that of intersectional feminist theory, which seeks to understand the marginalization of people according not just to gender but also sexual orientation, class, race, ethnicity, national heritage, and dis/ability. Thus, this chapter explores and applies a broad definition of feminist theory to editing practices, both in the classroom and the editing workplace, as one example of how theory can help to improve our work as editors and teachers.

First, a brief personal story: as a graduate student, I coedited a book about feminism with three other women, two of whom were also graduate students. In doing this coediting work, I also pondered the idea that feminism meant more than espousing feminist values and publishing works about feminism, but also that it might change my practices. To me, work was work, and I approached my assigned editing tasks in a very pragmatic, efficient, get-the-job-done manner. I learned to work from my grandfather, who held a very Puritan kind of work ethic. "Put more elbow grease into washing those windows, Susie; you're leaving streaks on the glass," he said repeatedly as I washed the grimy windows of his machine shop with newspaper (the irony of connecting the work of editing with the work of crumpling newspaper to wipe away window grime is not lost on me). After a couple of summers of washing his windows and not getting paid because the windows were still streaky, I learned to attack such jobs and other chores with more vigor and muscle, although with little finesse. I realized that I had applied this same work ethic to my editing tasks and perhaps to my collaborative relationships; I had valued getting the job done correctly the first time—the end-result—more than I had valued the people involved, to the detriment of what had been very good friendships and productive academic lessons. And I realized that perhaps feminism theories, which I had espoused before, could be more fruitfully applied to editing processes.

In this chapter I explore and interrogate the various scholarships that align editing with theory. Making clear one's theoretical foundation is both part of the argument—that we should recognize and understand the theoretical basis of our practices—and method—that understanding our theories helps us improve our practices and processes. I review the scholarship surrounding the topics of feminism, editing, and teaching. In reviewing the scholarship, I read deeply while questioning what each source can tell us about the unique practices of female editors, about the relationship of female editors to authors, to texts, and to readerships, about the enactment of feminist principles in editing workplaces, and, finally, about the enactment of feminist principles in editing classrooms. These questions, much like the research questions guiding empirical research, unearth the answers often buried in prior scholarship and here can help to illuminate the theoretical principles of editing practices.

While myriad works of scholarship recognize the role of female editors, some of which are explored below, few scholars have attempted to promote a theory of feminist editing, none recognize the role of feminism in editing workplaces, and none bridge the gap between teaching editing and feminist pedagogies. For decades, many women and feminist scholars have been speaking up to bring attention to gender equality, gender discrimination, and gender harassment in the workplace, but not the direct connection between feminism and editing. It is time to recognize and call attention to the important role that women play in the field of editing and in shaping editing practices and pedagogical approaches. This chapter attempts to rectify the feminist-editing gap by highlighting three avenues by which feminist theory can beneficially influence the editing classroom: as a focus on workplace practices, as a textual focus to approach the editing tasks so as to bring the author's meaning and audience expectations to full fruition, and finally as a pedagogical focus by which teachers interact with students.

These three strands are necessarily entwined; implementing one approach without another creates imbalance in the classroom, and leaves students confused and questioning the value of feminist pedagogies and feminist editing. Like Frost's argument (2014) that teachers of technical writing should direct students to engage with feminist theories, we also argue that technical editing classrooms do the same: as Frost states, "student engagement with feminist perspectives can aid in the development of new strategies for effective technical communication for a wide range of audiences" and such feminist perspectives "encourage thinking about the subjectivity of technical documents [and] textual production" (pp. 113–114). Viewing editing practices and pedagogies through a feminist lens is important if the field of technical editing is to understand and intervene in gender imbalances in technical communication and editing workplaces. I argue that understanding how gender imbalances affect editing workplaces is an important consideration for students of editing, one that is best learned through a direct and apparent consideration of feminist rhetorical theories in the editing classroom.

## Women in Editing Workplaces

Historically, women moved into editor positions in the field of literary studies and academic publication around the same time that suffragette movements were gaining ground. Indeed, as Fiamengo argues (2008), it was the suffragette movement that prompted much of women's writing in magazine, newspapers, and journals of the mid- to late nineteenth century. She specifically notes the work of Sara Jeannette Duncan, the first full-time female newspaper employee in Canada, after having left her position as literary editor of *The Washington Post*; at the Canadian paper, *The Globe*, she continued to participate in the editing staff. Duncan's treatises about the work of women's rights were witty, satiric, scornful of many other suffragettes, rhetorically savvy, and drew a large readership, to the point that she was often the topic of reporting by other journalists. While Fiamengo specifically notes the relationship between women's rights and women writers in the public sphere, she does not, however, describe the editing of such treatises nor the relationship between female authors, like Duncan, and editors. It seems clear, however, that Duncan's work as literary editor probably helped to establish her reputation as an intellectual critic and intelligent, rhetorically adept writer.

Scholarship in literary studies has also noted the increasing role of female editors during the suffragette movement. Irvine (2008) has traced the work of female editors, while describing the challenges they faced in working in realms more often filled by men. In his study of nonmainstream female editors in Canada during the early twentieth century, Irvine notes that women, despite their rhetorical adroitness and passionate entreaties for women's rights, were often denied positions of editorial responsibilities at that time. He describes the time as one in which mainstream publications were largely closed to women's publication: "That Canadian women seized that opportunity to edit modernist and leftist little magazines is in part attributable to the exclusion from editorial position in the mass-circulation magazine's normative, middle-class, male-produced, dominant culture" (p. 5). As a result of this exclusion, he argues, women who desired an opportunity to voice their theories and concerns turned to avant-garde, little-magazine editing and publishing, most times unpaid and voluntarily, to promote their political views and their poetry. In so doing, Irvine argues, such editors did much of the work typically assigned to a much larger team of editorial staff and supervisors:

> [T]he women poets and magazine editors . . . conducted promotional tours, solicited and collected subscriptions, courted advertisers, typed stencils, cut and pasted dummies, answered correspondence, and so on. Most little magazines routinely employed women . . . in clerical roles, sometimes acknowledged on the masthead, often not.
>
> (p. 17)

Irvine's study makes clear that, more than a hundred years ago, women were fighting for avenues to give voice to their theories and ideals, and if such avenues were closed to them in mainstream publications, they would go to extraordinary lengths to ensure that their own voices as well as those of other women were published, read, and followed by a readership equally hungry for such voices and views.

Then, as perhaps now, female editors saw their work not only as that of grammar marm, but more importantly as that of a deeper calling and a more meaningful engagement with authors, readers, and idealism. Irvine also notes that much of the editorial work of these pioneering women editors was clerical in nature, and thus often demeaned by other scholars and editors:

> Though crucial to the little magazine's non-commercial economy, these menial jobs have regularly been deemed inferior to editorial work and summarily disregarded by little-magazine historians. The predominantly clerical labour involved in little-magazine production is, in this histori-cal context, gendered female; this characterization, too, demonstrates that literary history has so far marginalized not only women editors and mem-bers of magazine groups but also their feminized forms of labour.
>
> (p. 17)

This gendered stratification of the status of various types of editing work is as noteworthy now as it was over a hundred years ago.

Irvine's study of avant-garde female editors of little magazines shows how strongly and how directly theory guided the practices of these women; they were deeply committed to the ideals of equality, of work, of socialism, of freedom of speech and of press. They cared about and cared for the relationship of author, audience, editor and text; they either subconsciously or consciously recognized the role of rhetoric in bringing discourse to their readership, and they absolutely were politically motivated to improve the lives of women through their work as editors. These were not workplaces devoid of theoretical implications; instead the editing workplaces of these women were highly and consciously embedded in several different and entwined theories: rhetoric, feminism, equality, and freedom. Irvine's scholarship also shows the very real struggles of inequality, unequal pay, discrimination, and disregard that such women faced in their editing and publishing practices.

Despite the lack of feminist theoretical approaches to editing scholarship, we recognize that gender imbalances occur, and probably more frequently than we care to publicize, in editing workplaces. One of the most popular and publicly available editing sites, Wikipedia, reports wide gender gaps, with male writers and editors much more likely to be published. Hargittai and Shaw (2015) report what is widely known in social media circles: "[G]ender disparities in who authors the material on the site are increasingly a matter of public concern as the content disproportionately reflects the interests and perspectives of its most active contributors who are mainly men" (p. 424). And they note that previous campaigns by the Wikimedia Foundation to promote greater balance of contributors between the genders proved largely unsuccessful, even after more than three years. Their study strongly implicates the imbalance of technological competence between women and men, who they argue are more likely to be trained in computer coding and programming skills. Their study suggests that men are more likely to seek editing participation in such sites as Wikipedia than women because of a significant internet-skills gap between men and women. The gap in internet and other digital skills may be explained by more than a simple correlation to gender, as Hawisher and Selfe (2004) assert. They describe a strong and continued tradition in English classrooms of emphasizing written text over digital, multimodal compositions:

> [I]n English composition classes and in most official assessments of communication ability, the focus is primarily on conventional alphabetic and print literacy. To a large extent, this valuing of official forms of alphabetic and print literacy is generational. Such literacies have, after all, been the major shaping forces in the educational experiences of faculty members teaching at these schools and thus in the ongoing formulation of their official grading and evaluation standards. . . . Raised and educated in a culture that valued, and continues to value, alphabetic and print literacies, many instructors remain unsure of how to value new-media literacies, unsure how to practice these new literacies themselves, and unprepared to integrate them at curricular and intellectual levels appropriate for young

people. In this context, instructors cannot take full advantage of the literacy strengths computer-savvy kids bring to the classroom and may miss important opportunities to link their own instructional goals to the developing literacy strengths of talented students.

(p. 20)

Such classroom instruction, which de-emphasizes internet, digital composing skills largely as a result of and preference for the teachers' decades-old textual instruction, may also help to explain why a gap in internet skills continues to exert itself even in hyperdigital environments like Wikipedia.

In regards to the role of women in contemporary editing positions, Sheffield, Sparks, and Ianetta (2014) describe their collaborative work as editors of the proceedings of an academic symposium, and they note that the work of editing academic journals frequently falls to "apprenticing" graduate students, often likely to be women. In such work, these apprenticing graduate students are rarely recognized for their skill, but are often called on to accomplish the clerical work that will enhance the careers and reputations of more advanced scholars. Thus, such graduate students serve the needs of other scholars and authors, while developing professionalization skills that they believe will be valuable in their own careers, and such is often the case. The point in their description relevant to this study is that the majority of these editing apprentices are women, and that they perform and are perceived as servants rather than as professionals.

Similarly, the Bureau of Labor Statistics (2016b) reports in the "Current Population Survey #39" that, in 2016, 64,000 men were employed as editors, while 56,000 women were similarly employed. However, when digging a bit deeper into the data, we note that in the lower status job of copymarkers and proofreaders no men were reported to hold those positions, while as many as 5,000 women did. Such data suggest that men may be more likely to be promoted from the proofreader stage, but that women stay in such lower-level positions longer. The data also suggest that men may be hired initially at positions of higher responsibility and status than women. Alternatively, the data suggest that women may prefer entry-level positions like copymarkers and proofreaders, perhaps when reentering the workforce or while working part-time. A separate statistical survey by the Bureau of Labor, also in 2016 (CPS #11, 2016a), reports that 52% of the total editor workforce were women. Regardless of which percentage is correct, it is clear that near or slightly more than half of the editing workforce is staffed by women, but that women are more likely to be employed in the entry-level positions of copymarkers and proofreaders, the clerical strata, than men.

## Feminist Approaches to Editing

Feminist theories in editing go much further than simply addressing gender inequities in the context in which editing exists—the editing workplaces.

Feminist theoretical approaches can also illuminate and influence editing practices; that is, feminist theory can help us understand how we work with texts and authors.

Traditional approaches to editing practices assumed that editing practices were devoid of theory, an assumption that also undergirds a view of technical discourse as objective, distanced, and theoretically neutral. However, when we hold Buehler's (1980/2003) assertion that using such language is a theoretical choice (for her, rhetorical theory), then we also believe that editing practices which develop such "neutral" style are similarly theoretically entangled. A feminist theoretical approach to the text can shed light on how much theories of neutrality, tradition, mastery, correctness, efficiency, or other culturally bound theories subconsciously impact our practices. It is no less theoretical to assert that good editing should be done through mastery of grammatical norms and rules, what Hayhoe (2015) might regard as "a focus on microlevel topics such as grammar, syntax, and punctuation, the areas where our students are weakest" (p. 351), than it is overly theoretical to claim that feminism can beneficially shift how we perceive and respond to texts. Not all of feminist theory will be relevant to textual practices, of course, but the values of equality, respect, and awareness of power structures—brought to the forefront in much of feminism—can certainly reshape our textual practices as editors.

One such editor, Catherine Stimpson, revealed as much in her commentary on her five years spent editing the feminist, interdisciplinary journal *Signs*. In this autobiographical essay, Stimpson (1979) describes her work publishing the journal, and the work of the other women helping to edit the journal since its inception in 1974, as a balancing act between avant-garde radicalism and a "faithful[ness] to many rules that govern the contemporary academic text" (p. 37). Of their publishing vision and decisions, she explains: "We were trying to make our radical transgressions more palatable through making our acknowledgement of existing methods palpable" (p. 37), careful not to alienate too many readers and administrators in being too transformative or too interdisciplinary, one of several criticisms cast at the journal. She continues, describing the dismissive reception that many academics held about such interdisciplinary explorations:

> Young scholars have told me that it is bad enough to write about women, let alone to publish in an interdisciplinary journal. If they do so, they will be considered "soft," a metaphor for professional inadequacy that casts "hardness" as a metaphor for professional virtue.
>
> (p. 38)

In these remarks and in what they represent of her work, Stimpson reveals an attention—an awareness of rhetorical theory—to both the work of authors and the importance that such work plays in their careers, as well as to the role of audiences in helping to shape and mold the journal. She is attuned to the needs of both author and audience, and in helping to craft the journal to those expectations. Such awareness is, of course, rhetorical, but it also suggests a feminist work mode that respects the author and the audience equally to the goals and vision of the journal and the work of the editor.

In a more deeply feminist mien, Stimpson (1979) also describes specific ways of relating to the texts and textual practices: "We extensively referee articles and intensively edit them. . . . We jostle ideas against each other, as if juxtaposition and exposure had the power of fresh air in a dank room" (p. 41). These processes of refereeing articles, intensive editing, and a dialoging about ideas are not unique to feminism; these processes existed in editing long before the advent of feminist theory. Yet, here, when viewed in a feminist lens, they are given renewed meaning, a meaning that is thoroughly implicated in a theory of textual and authorial respect and equality. In other words, while traditional modes (perhaps masculinist modes) of editing also use these same processes, traditional editing does so as a way of maintaining mastery and control over the style of a text. Traditionally, editing is less about being fully aware of one's position to the text and the author, and more about whipping the text into publishable shape. In opposition to this traditional value, Stimpson articulates an editorial sense of equality and respect:

> Our office has tried to embody certain feminist and egalitarian principles. Though we judge the competence of the 500 or so manuscripts we get each year, we try to strip the trappings of hierarchy from the process. We want to create a sense of participating in a shared, fair process.
>
> (p. 40)

Here, we see that a feminist theory can directly change specific editing practices: that is, sharing the text, being fair to the words and ideas of the author, and respecting the views of other editorial staff.

Perhaps more than any other specific feminist editorial practice is the use of feminine metaphors to describe the work of editors, metaphors that bring to mind the very womanly work of motherhood and midwifery: such verbs as "bring forth," "deliver," "nourish," "enable," and "nurture" abound in several editing treatises. In her survey of reviewers and the work of reviewing, Wisker (2013) describes the work of editing as

> the developmental work of reviewers and editors in *nurturing* and *ensuring* the quality of academic writing before it reaches publication contributes to dialogue in the academic community, one which we produce and spread ourselves and a practice which we both *nurture* and *safeguard*.
>
> (italics added for emphasis, p. 347)

Another female editor, Amy Einsohn (2011), also uses a maternal metaphor in attesting to the need for respect and tact when working with authors:

> Another approach to query writing is to treat the manuscript, no matter how poorly written or prepared, as though it were the author's ugly newborn. . . . You don't have to coo over a manuscript, but you should remember that it is the product of the author's labor and sweat, hopes and dreams.
>
> (p. 45)

In highlighting the metaphoric discourse of editing, Stimpson (1979) describes the language as "familial" and "parental" (p. 39), and later describes her own place in editing as "a calling" (p. 42). Clearly, such metaphors are necessarily not limited to female scholars, writers, and editors. Indeed, male editors may also use such feminine metaphors to describe their work. However, such metaphors, based on traditional women's work as mother, familial caretaker and nurturer, committed to the protection and wellbeing of one's children, are based on a view of one's relationship to texts and to authors—as caretaker, nurturer of words and meanings, protector of others' feelings, as a careful deliverer, like a midwife, of one's oeuvre to the audience.

Other specific textual practices that might arise from a feminist mode of editing include an awareness of the role that emotion and empathy can play in editing, that is, a distinct turn away from a dispassionate, impartial façade. Not all emotions are feminine, and not all feminist theory equates feminism with emotion; however, much feminist theory does emphasize the importance of emotion and does decry "pure" impartial logic as patriarchal trappings. In fact, many feminist scholars do value and call attention to the importance of emotion in scholarship. Emotion and passion are clearly evident in Stimpson's essay, as well as in Irvine's study of female editors of little magazines. Stimpson admits to conflicting emotions and a growing awareness of the power structures built into the editorial and publication structures that give rise to conflicting emotions in her position as editor:

> Editing *Signs*, I discovered how much I preferred saying "yes" to saying "no"; how much happier I was taking an article than refusing it. Indeed, perhaps all editors can be divided into those who sublimate a desire to give into their work and those who sublimate a desire to deny. Both acts, of course, signify forms of power. . . . Talking about power discomforts me.
>
> (pp. 39–40)

Given the passion and zeal that these female editors bring to their work, it is easy to see how such emotion can be beneficial to editing. While many scholars would claim, and indeed both Wisker and Stimpson do, that such zeal might hinder the careful diligence of intensive editing and editorial decision-making, I would argue that such emotion and passion would just as likely keep an editor committed to the work of publishing and in developing an empathy for the author and the text. Bored, dispassionate editors who are uninvested and uninterested in the manuscript or the topic would be as equally likely to warp and hinder a text's improvement as a zealous editor would be in overlooking a manuscript's logical gaps and errors. Some passion and emotion can be beneficial to the editing enterprise, so long as that emotion is recognized and thoughtfully managed, a cognizance made more likely through a clear approach of feminist theory.

Feminist theories—specifically those of intersectional feminism, which attends to how people are often marginalized because of gender, as well as sexual orientation, ability, ethnicity, national heritage, class, and religion—might also work to improve publication and delivery processes by attending to how marginalized people can access edited and published texts without undue prohibitions of cost and opportunity. Feminist editors might ask, "How can this text be made more readily available to readers without access to academic libraries?" and "How can this text be made more welcoming to readers who struggle with physical abilities of eyesight and cognition?" These questions and others like them that pay attention to issues of publication and/or mode of delivery can very likely be answered in beneficial ways with an overt attention to theories of equality and respect.

Entwined with threads of feminism in the context of editing workplaces and in regard to specific editing practices of caring for the text and the author, a third strand of feminism as it relates to pedagogical practices is also necessary for a full picture of how such theory can enhance the profession of editing. Most academics interested in scholarship pertaining to editing, as most authors in this book are, are selected to teach editing. In fact, many published pieces on editing invoke a call to teaching, as both Warren (2010) and Hayhoe (2015) have done in their articles. Hayhoe goes so far as to argue that, as a field, technical communication should probably be teaching more editing than most programs currently engage. In short, as much as we edit, we also teach editing for a future generation of technical communicators and editors.

## Feminist Pedagogical Practices

Frost (2014) has called for the clear and focused teaching of feminism in technical communication courses, an approach she terms "apparent feminist pedagogies." In her course, she makes transparent her own feminist ideologies and asks students to research and interrogate culturally loaded terms like "objectivity" and "efficiency." She asserts that apparent feminist approaches in the classroom offer several benefits, including drawing "attention to the fallacy of pedagogical objectivity," of "objectivity of fields of knowledge," and of "subjectivities in classroom dynamics" (p. 114). In her empirical study she describes several challenges that both she and the students encountered while teaching such a class, yet she maintains that making such theoretical approaches apparent to students is highly beneficial in helping students understand the ideologies of their major fields of study and in prompting students to "engage with cultural studies and social justice; in so doing they would come to a more critical understanding of what is happening—who is being marginalized—when the term *objective* is invoked" (italics in the original, p. 115). For Frost, in a technical rhetorics course, both the theory chosen (i.e., feminism) and the "apparency" (I know; this is an awkward nominalization of an adjective, but I don't think Frost would agree to "transparency") are equally necessary for student growth and learning. She outlines several different pedagogical methods by which she employed apparent feminism in her course: acknowledging her identification as a feminist, assigning and acknowledging the feminist theory she would ask students to research and analyze, providing technical documents for their theoretical analysis, and requiring students to use written responses as a place to develop thoughtful, productive responses to questions that prompted critical thinking. To be clear, Frost did not require nor expect students to align their beliefs with feminism; she only asked that they use such a theory in their cultural analyses in the hope that they might learn to critically analyze other cultural hegemonies. While she did experience some resistance from several students—students resistant to both feminism and to acknowledging other forms of hegemony—she also saw much growth of critical thinking skills and social justice awareness in many students.

## Pedagogical Applications

Like Frost did in a technical communications course, feminist pedagogies can be brought to bear on editing courses, although perhaps with different end-goals. For Frost's class, goals of promoting intradisciplinary knowledge and of promoting social justice were preeminent. For an editing course using an apparent feminist theory, the goals may be less focused on social justice and analyzing whole-scale hegemonic cultural values, and more focused on building awareness of the theories that underlie all practices, including editing practices, and awareness of the power structures inherent in mediating between author and audience, text and meaning. While altering the end-goals may be prudent, some of Frost's pedagogical methods may be very productive in an editing classroom: making one's values apparent to students; making a theoretical basis, here feminism, clear and accessible to students, providing them with samples for skill application, asking them to apply critical analysis in thoughtful, well-developed responses. As Frost experienced, so we may also see some resistance from students who, like so many other well-versed technical communicators, are deeply committed to perceiving technical discourse as neutral and objective, and we may see resistance from students who view the work of editors as mere practice rather than as theoretically embedded habits. Despite such resistance, however, we would do our students greater service and benefit in openly acknowledging a feminist theoretical approach, than in denying any kind of theoretical basis.

In teaching editing as a feminist practice, we would help students see the subconscious values that they might bring to the editing classroom, for example, values such as efficiency, profit-building capitalism, impartiality, neutral transmission of meaning, master of correctness, among others. We would also help students recognize other theories that could be equally productive—perhaps poststructuralist, postcolonialist, or actor-network theory, which I do not propose to describe or apply, however briefly, here. Students often surprise me with their insights, and I do not doubt that there are many students, who once prompted, could make ample and reasonable arguments for other theories that would prove fruitful in analyzing editing practices. I welcome such insights from my classes.

Further, in teaching editing as feminist practice, we might encourage greater affinity for textual adroitness and empathy for authors among our students. In describing and applying strategies to textual practices that move beyond sheer correctness and efficiency, we can do both our students and our manuscripts a great service. Most manuscripts and most students need time and thoughtful consideration to develop and improve; we need to teach our students deep, close, and slow reading skills in order to teach them how to work with a text to reach its full potential. Such reading skills are quite productively taught by using a feminist theory that highlights an affective affinity to the text and a respectful empathy for the author and audience. When students learn about feminism, they can also learn to value and respect the work and words of authors, and to approach the text not as a chore that has to be completed, but rather as a product to be brought to fruition. Without explicitly describing a feminist theoretical approach to textual practices, Ball (2018) does describe on her website her editing classroom practices as having similar qualities, as she relates teaching students how to balance the nuances of respect for all the stakeholders of an editing task:

> We discussed how they might honor the author's voice (the concern that kept them from working more substantially with the text) while also fulfilling their responsibility to the venue, its readers, its design constraints, deadlines, and budgets. The students learned that each editorial decision is a delicate rhetorical act within a larger ecology, not just a contextual line-by-line correction of comma splices.

In Ball's classroom, as in her work as editor of digital, multimodal texts, she works with theories that are attuned to the nuances of the full context of any editing task. Likewise, an editing classroom that enacts feminist theoretical principles will also encourage collaboration, respect for all stakeholders, and situational rhetorical awareness; it will also likely help students recognize and expand beyond traditional editing hierarchies, moving beyond traditional separations of editor-in-chief from copyeditors. While there is no denying that such hierarchies do exist, and sometimes for good reason, a feminist intersectional editing classroom will encourage students to understand the hierarchical principles while also helping them to see the ways in which such hierarchies are fluid, subjective, and often uselessly enforced, moving the class instead to a more collaborative venture. While many apparent theories might prove productive in helping students recognize the subconscious values they bring to their practices, feminist theory with its frequent valuing of emotion and equality would be most beneficial in helping students respect and value a text, author(s), and audience(s).

I acknowledge that feminist theory will do little to change the grammatical rules and style guidelines that often form the bulk of work done to texts; in other words, a fully edited manuscript edited with feminist theory in mind may look much the same, perhaps exactly the same, as a manuscript with no apparent conscious theory. Yet, it is to be hoped that the feminist theory might alter the mindset by which copyeditors and grammatical hawks approach the task at hand. Further, it is likely that approaching the editing tasks with feminist theory will result in a greater appreciation for and relationship with the authors and the readers, as well as with other coworkers. Any theory that highlights and encourages equal respect among the various stakeholders—authors, readers, coworkers, supervisors, texts—will surely, at least some of the time, also promote a more equitable and respectful workplace and improved publishing process.

Finally, bringing feminist theory to the editing classroom can provide a realistic prospect for many of our students' future careers. As the labor statistics show, gender discrimination in editing workplaces can influence the lack of advancement of women into positions of higher status and greater responsibility, like managing editor, comprehensive editor, and editor-in-chief. As writers, reviewers, and editors, we would do well to work to promote greater equality among all employees in our editing workplaces and to recognize greater respect of women's abilities for advancement. As teachers, we should reveal to our students the realities of editing workplaces, both the rosy pictures of future career potential and pay growth as well as dismaying prospects of discriminatory practices. In doing so, we can hope that our editing students may work to promote and negotiate for greater equality and respect in their workplaces too.

## Pedagogical Practicalities

This chapter closes by reiterating several pedagogical methods in regards to feminism and editing.

- Make clear to students that all practice, including editing practices, are entangled with theory.
- Openly acknowledge the feminist approach that will be used in the classroom and declare the goals of that approach—to help students recognize

the often-subconscious theories that are culturally embedded in editing and to help students develop a greater appreciation and respect for texts, for authors, and for audience.

- Supply students with texts for feminist analysis, including but not limited to texts that call for gender neutral pronouns and other gender-specific examples.
- Allow time for and require students to respond thoughtfully and in well-developed responses to critical thinking questions.
- Prompt students to consider other theories that might help them productively work as editors.
- Acknowledge the important role of emotion and zeal in motivating editing and editorial work and guide students in learning to manage that emotion productively.
- Require students to work with real authors and real texts, so that they can come to learn how to respect the work and words of others, not just as a matter of correctness, but as a situation of cooperation, respect, and fairness.
- Supervise such student–author projects closely so that timely and appropriate guidance can be given when needed.
- Call attention to gendered discriminatory practices when noticed in the classroom and in hypothetical future workplaces.
- Help students develop strategies for improving editing workplaces in regards to gender discrimination and/or harassment.
- Recognize the subconscious theories and values that underlie one's own practices both as editor and as teacher.
- And, finally, learn to respect and appreciate the skills and efforts that students bring to our classrooms, treating them not quite as equals, but perhaps as developing partners in helping to bring forth improved manuscripts and publications.

## References

Ball, C. E. (2018). Philosophy: An editorial pedagogy. Retrieved June 30, 2018 from http://ceball.com/philosophy

Buehler, M. F. (2003). Situational editing: A rhetorical approach for the technical editor. *Technical Communication*, *50*(4), 458–464. Reprinted from 1980.

Bureau of Labor Statistics. (2016a). CPS #11. Employed persons by detailed occupation, sex, race, and Hispanic or Latino Ethnicity. Retrieved from https://www.bls.gov/cps/cpsaat11.htm

Bureau of Labor Statistics. (2016b). CPS #39. Household data annual averages: 2016. Retrieved from https://www.bls.gov/cps/cpsaat39.pdf

Dragga, S., & Gong, G. (1989). *Editing: The design of rhetoric*. Amityville, NY: Baywood.

Einsohn, A. (2011). *The copyeditor's handbook: A guide for book publishing and corporate communications*. Berkeley, CA: University of California Press.

Fiamengo, J. A. (2008). *The woman's page: Journalism and rhetoric in early Canada.* Toronto, ON: University of Toronto Press.

Frost, E. (2014). Apparent feminist pedagogies. *Programmatic Perspectives, 6*(1), 110–131.

Hargittai, E., & Shaw, A. (2015). Mind the skills gap: The role of internet know-how and gender in differentiated contributions to Wikipedia. *Information, Communication & Society, 18*(4), 424–442. Retrieved from http://www.webuse.org/p/a49

Hawisher, G. E., & Selfe, C. L. (2004). On editing and contributing to a field: The everyday work of editors. *Pedagogy, 4*(1), 9–26.

Hayhoe, G. (2015). Boom, bust, and beyond: My experience as a technical communicator. *Journal of Technical Writing and Communication, 45*(4), 342–353. doi:10.1177/0047281615585748

Irvine, D. (2008). *Editing modernity: Women and little-magazine cultures in Canada, 1916–1956.* Toronto, ON: University of Toronto Press.

Sheffield, J. P., Sparks, S. C., & Ianetta, M. (2014). Symposium: Revaluing the work of the editor. *College English, 77*(2), 146–164.

Stimpson, C. R. (1979). Editing "Signs." *The Bulletin of the Midwest Modern Language Association, 12*(1), 37–42.

Warren, T. L. (2010). History and trends in technical editing. In A. J. Murphy (Ed.), *New perspectives on technical editing.* Amityville, NY: Baywood. doi:10.2190/NPOC3

Wisker, G. (2013). Articulate—academic writing, refereeing, editing and publishing our work in learning, teaching and education development. *Innovations in Education and Teaching International, 50*(4), 344–356. doi:10.1080/14703297.2013.839337

# 6

# EDITING FOR HUMAN–INFORMATION INTERACTION

*Michael J. Albers*

## Chapter Takeaways

- Comprehensive editing requires editing for high-level organizational and audience issues to answer the basic question of "does this document communicate its information efficiently and effectively?" It is important that editing courses teach students how to perform this level of edit.
- The pedagogy of teaching developmental editing needs to include teaching how to analyze documents at a global level.
- A study on graduate students' comments found the students were competent at both sentence and paragraph level editing, but improvements are needed on editing for global issues.

## Introduction

People come to a text because they want to use the information in some way. How well they can use the text depends on how effectively it communicates. Thus, both authors and editors have to consider why the information is relevant to the reader and how to best present it to maximize the communication. Both must be concerned with human–information interaction (Albers, 2012a). Editing for human–information interaction requires an understanding of the audience and how the audience will use the information. This chapter explores the need for editing to focus on how people interact with and use information, which is more specific than the high- and low-level textual editing typically taught. More specifically, it raises questions about what is meant by high-level editing and considers how to better address it in the classroom.

Complex information that is typically of concern when dealing with human–information interaction is used for decision-making and problem-solving.

That factor of needing information forms the basic definition of human–information interaction.

> HII [human–information interaction] starts with people. HII deals with the abilities and nature of a person—what happens in the mind—as it covers how people find, interpret, and use information to best achieve their goals . . . HII deals with understanding people, and how we use that understanding to improve the behavior and overall usefulness and desirability of systems which provide them information.
>
> (Albers, 2008, n.p.)

It also involves realizing that a failure of information communication may arise because of the failure to understand how the information is communicated and how people will interact with and interpret it. As worded, that sentence seems to specifically address authors, but editors also bear responsibility for ensuring a text communicates.

Performing a comprehensive edit could be thought of as making revision decisions about a text. But this brings up the question of what is meant by "revision decisions?" Within this chapter—and what should be a major aspect of an editing course—these revision decisions should be based on the human–information interaction elements. A comprehensive edit needs to address what information a person needs, how they need it presented, and how they will be interacting with it. Thus, an editor faces two challenges: first, the human–information interaction problem must be identified and, second, the problem must be communicated to the author. This chapter does a preliminary examination of whether student editors identify audience and make comments about human–information interaction. It is unrealistic to expect students to learn how to edit at any level without explicit instruction, so we may also have to make pedagogical changes to enhance teaching editing for human–information interaction.

This chapter begins with a short literature review on high-level editing and writing editorial comments. It then examines the editorial comments on two different graduate course editing assignments. Finally, it explores the results and considers the implications for editing course pedagogy.

## Literature Review

We can, at a high level, define an editor's job as ensuring that the writer's view (and, consequently, the text's view) of the content matches the reader's view to ensure high quality communication of the text's information. Gulliksen and Lantz (2003) point out that "communication is identified as one of the key issues that needs to be addressed to achieve well-functioning user-centered design" (p. 5). Developing a text that provides that information view in an effective manner requires interacting with an editor who works at the comprehensive and organizational level (Willen, 2004). Technical text creation requires a joint effort between the writer and the editor, with the editor tasked with verifying that the text fits the reader's information needs and providing the writer with "ways to make the document easier for readers to understand and use" (Rude, 2006, p. 12).

Addressing technical communication in general, Hayhoe (2002) said, "One of the most crucial tasks of the technical communicator is to provide information that users need by carefully selecting the right mix of content and then developing, arranging, and presenting it effectively for the audience" (p. 398). An editor provides valuable assistance in developing, arranging, and presenting that information effectively. It would be easy to argue that, with respect to editing, communication with a goal of maximizing comprehension is the most important issue. Thus, one aspect of teaching comprehensive editing requires teaching students how to edit for human–information interaction. Learning to edit requires developing the ability to spot major problems within the text. This is a major skill that most editing texts refer to as either substantive or comprehensive editing. Regardless of the term used, this editing process refers to the overall text structure and is not a focus on the grammar or syntax of the text.

Past work has found that authors want good comprehensive editing. Walkowski (1991) and Winsor (1993) both studied software engineers and found that they wanted more than a basic copyedit. They expected comments on how to improve the text, including reorganization. This finding is a contradiction to the potentially antagonistic editor–author relationships found in many editing texts, but is in line with Lanier's (2004) findings. Albers and Marsella (2011) found that most students grasped the basic concept of how to perform a comprehensive edit and could make solid constructive comments. Ignoring the few students who tried to rewrite rather than edit the text, most produced a consistently high number of comments that were basically of good quality. However, they also found that students made few global-level comments and, rather, focused on paragraph-level issues. Global-level problems are the larger organizational problems that impair human–information interaction and identifying them should be a major goal of a comprehensive edit.

If students understand how to perform a global-level edit, then multiple editors working on the same document should all produce a relatively consistent set of edits. At the very least, most (if not all) of the major problems within the document should be marked. That situation, however, was not evidenced in Albers and Marsella's study.

Creating a text appropriate for an audience is a standard tenet of technical communication. Obviously, this must apply to editorial work too. Dragga and Gong (1989) argue that editors need to address the real audience, rather than any abstract readership. This simple-sounding aspect of writing is actually very complex and can be a major stumbling block for student editors, whose view of audience tends to revolve around an audience of one, the instructor. Buehler (2003) considers what she calls a "programmatic approach" to editing dominated by a knowledge of grammar and other mechanics of spelling and punctuation, which she acknowledges is necessary, but not sufficient. What is missing in a "programmatic approach" is the focus on higher-level document structure and audience needs.

Essentially every modern technical communication textbook has an early chapter devoted to understanding the audience. They tend to fall into describing that audience as a demographic group with needs, but avoid addressing how human psychology drives the information interaction and comprehension. Yet, these are the essential elements that a writer and editor must know in order to communicate high quality information. Likewise, editing textbooks discuss various levels of editing, ranging from the word/sentence level to the document level. Ensuring a text effectively communicates requires thinking of the text at the organizational level. I'll admit that a text rife with grammatical errors does impair its communication ability; however, within a typical corporate environment—in which we are teaching students to operate—this should not be a significant problem. A corporate manager will not approve releasing a text with a high number of basic errors. On the other hand, with just copyediting and not quality comprehensive editing, it is very easy to release a text that is grammatically perfect and well formatted but that totally fails to communicate its message. All of the needed information may be within the text, but poor organization can derail a reader's ability to interact with that information.

Thus, we are forced to question whether student editors really consider audience and apply those considerations to a text. Audience is not just demographics or just a group of readers; it is a group sharing information needs and text interactions. Are students learning how to consider audience or is it something to nod about during a lecture and then ignore when they apply red pen to paper (red text to file just doesn't have the same ring)? Unless the basic concept of meeting audience needs is deeply ingrained, then editing for human–information interaction will not occur.

Most comprehensive editing is done via comments rather than direct text changes. These edits call for more than single-word or grammatical changes and require the author to make those changes. As a substantial element in comprehensive editing, comment writing has received significant attention. Developing comments has been described as a multilevel process (Anderson, Campbell, Hindle, Price, & Scasny, 1998; Van Buren & Buehler, 1980). Reflecting this multilevel process, editing textbooks call for multiple passes through a text, starting with the high-level structure and working down (Rude, 2006; Samson, 1993).

Clearly, a comprehensive edit depends on the quality of the comments, and Eaton, Brewer, Portewig, Davidson, and Portewig (2008) found that about 72% of comprehensive edit comments are followed. If comprehensive editing comments are followed, then editing course pedagogy must stress developing high quality comments. Albers and Marsella (2011) looked at undergraduate commenting style and the use of direct and indirect comments. Although well formed, the comments were not as useful as they could be since they focused on lower-level aspects of the text. It seems we need both more research into developing comments themselves and into how to best analyze a text as the preliminary step before making the comments (Albers, 2012b).

## Methods

Two assignments in my graduate editing classes have students edit two recommendation reports: one evaluates children's software and the other evaluates three copiers for an office. The assignments were given on consecutive weeks toward the end of the comprehensive editing units, which followed a copyedit unit.

The editing directions given on each assignment were as follows.

Software recommendation report (1800 words)

Perform a comprehensive edit. In your edit, recommend headings, lists, graphics, and tables to improve the overall look and readability of this report. If you see comprehensive-edit problems, bring them to the authors' attention. The audience of this letter is the kindergartener's parents. If you see discussion that may be unclear for these readers, include a note to the authors or suggest revisions if you can.

Copier evaluation report (1000 words)

Perform a comprehensive edit, write a letter of transmittal, and create a style sheet. In your edit, recommend headings, lists, graphics, and tables to improve the overall look and readability of this report. If you see comprehensive-edit problems, bring them to the authors' attention. The audience of this letter is the managers responsible for deciding on copier purchases. If you see discussion that may be unclear for these readers, include a note to the authors or suggest revisions if you can.

All three classes performed the software assignment then the copier assignment on consecutive weeks.

Both assignments had multiple serious high-level and low-level problems. I don't view them as a realistic editing, but rather as closer to the texts with horrendous copyedit errors that are used as the early assignments in copyediting.

The papers analyzed for the study came from three different sections of a graduate editing course, taught in different semesters. The East Carolina University IRB approved the study. A total of 60 papers were available for coding and 56 papers were coded. Each comment was coded as either global or non-global.

Four papers of the original set were not coded. They received C and D grades and those student-editors did not perform a comprehensive edit, but rather performed a light copyedit and/or a partial rewrite of the document. They were not coded because (1) they contained minimal comments at any level and none were global and (2) with only four papers, the low-grade papers were too few for meaningful analysis.

## Results

In the 56 papers, a total of 1567 comments were evaluated and 210 were coded as global.

The first analysis round looked at the comments versus the grades the student received. The analysis used an A/B grade breakdown. There were not enough A-, B+, and B- grades to use five groupings in the statistical analysis. Table 6.1 shows the grade breakdown. The numbers for the two assignments do not match because there were a different number of papers submitted.

A summary of the descriptive statistics appears in Table 6.2. We can see wide variations in the differences between the minimum and maximum number of comments. The high maximum for total comments came from a couple of editors who could be accused of over-commenting. Many of these comments were explaining copyedits, such as "a comma was missing" or "changed verb tense," which were consistently edited in-line but not commented on by other editors.

### Analysis of Total Comments

The total number of comments across the two papers appears similar (Figure 6.1).

Interestingly, it almost appears to be a bi-modal distribution, but there is not enough data to determine if the 40–45 comments are actually a second peak. A bi-modal curve could be caused by the results coming from two populations with, in this case, different commenting styles. A normal curve for total number of comments would be a reasonable expectation. On the other hand, the potential bi-modal distribution makes sense because the editors who made more than 45 comments could be accused of over-commenting. The distribution from a larger sample—and a sufficiently large sample including professional editors—would be interesting to determine if the bi-modal distribution is real or an artifact of this dataset. The bigger study would reveal if we should be trying to progress student editors to the second (higher number) peak.

A graph of the number of global comments made for each paper also appears similar (Figure 6.2). An interesting observation is the lack of high-end outliers.

**TABLE 6.1** Grade breakdown of the copier and software editing assignments.

|  | A papers | B papers |
|---|---|---|
| Copier | 84 | 18 |
| Software | 80 | 32 |

**TABLE 6.2** Descriptive statistics of the comments made on the copier and software editing assignments.

|          |         | Total comments | | | Global comments | | |
|----------|---------|-----|-----|--------|-----|-----|--------|
|          |         | min | max | median | min | max | median |
| Copier   | overall | 4   | 87  | 22     | 1   | 7   | 4      |
|          | A       | 11  | 87  | 23     | 2   | 7   | 5      |
|          | B       | 4   | 45  | 18.5   | 1   | 4   | 2.5    |
| Software | overall | 7   | 80  | 24     | 0   | 7   | 3      |
|          | A       | 15  | 80  | 34     | 2   | 7   | 5      |
|          | B       | 7   | 39  | 21     | 0   | 7   | 2      |

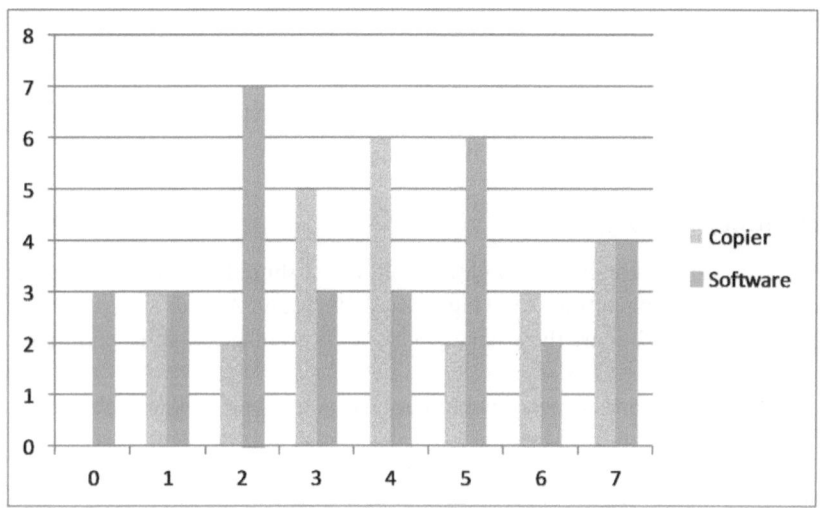

**FIGURE 6.1** Total comments for the copier and software reports.

Even the editors who over-commented on the text did not make an excess of comments at the global level.

The total comments made for the copier and software assignment two-tailed t-test gave $p = .450$. (The level of significance for this study was $p \leq .05$.) Likewise, total global comments made for the copier and software assignment two-tailed t-test gave $p = .288$. This implies they are from the same population. Thus, we can conclude that the difference in the total number of comments is not significant and we can combine the two sets for further analysis of total comments.

Looking at the differences between the number of comments for A and B papers a two-tailed t-test found: total comments $p = .002$ and global comments $p < .001$. Not a surprising finding that A students produced both more comments and more global comments than B students.

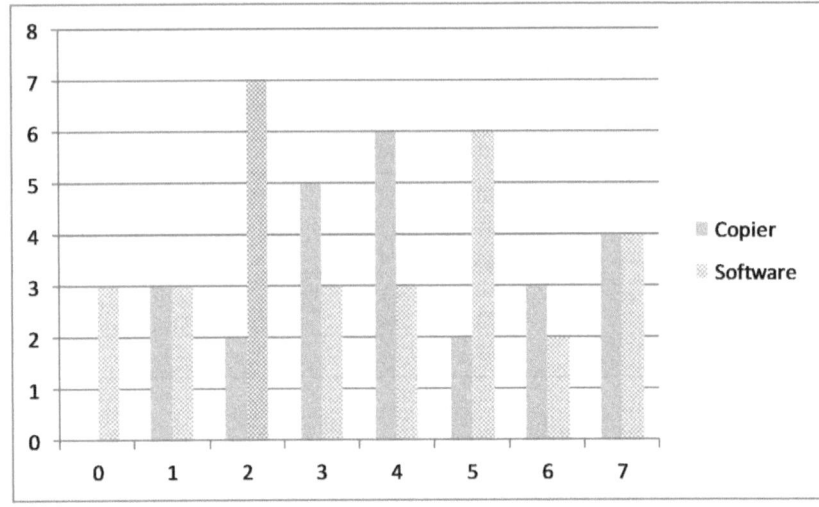

**FIGURE 6.2** Global comments for the copier and software reports.

## Percentage Analysis

The ratio of total comments to global comments should be more interesting than simply looking at the totals (Figure 6.3). We found A students made more global comments, but are they making a higher percentage of global comments (Figure 6.4)?

Both the copier and software assignments show a strong peak with about 20% of the total comments being global.

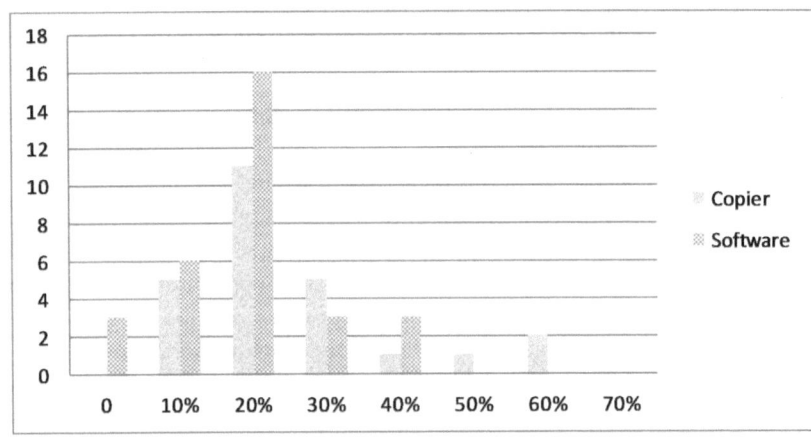

**FIGURE 6.3** Percentage of global comments for copier and software assignments.

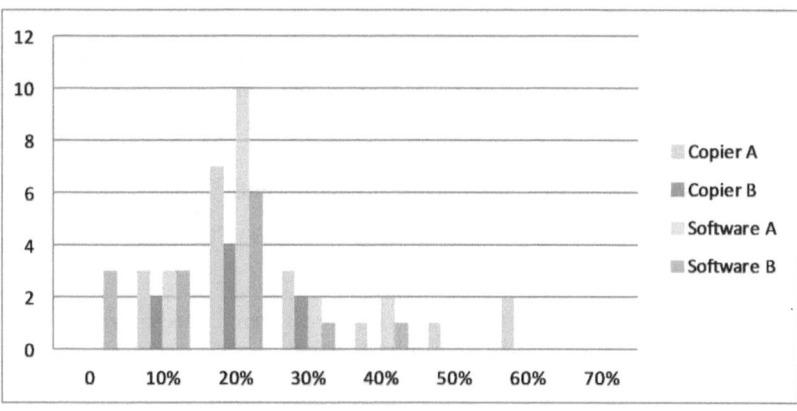

**FIGURE 6.4** Percentage of global comments by grade.

Looking at the by-grade breakdown the percentage of global comments retains the strong peak at 20% and, as would be expected, that at higher percentages, the percentage of global comments by A students was greater. For comparing Figure 6.3 and Figure 6.4, the height of the A/B bars in Figure 6.4 is equal to the bar in Figure 6.3.

An analysis of overall percentages of copier versus software two-tailed t-test gave $p = .049$. In other words, when we looked at the percentage of global comments, the copier report and the software report showed a statistically significant difference. A somewhat unexpected and undesired finding. This implies that they are from different populations. Looking at Figure 6.3, we can see it is probably the result of the long tail of copier comments that the software comments does not have. Unfortunately, it draws into question whether the copier and software global comment percentage values should be combined for the further analysis.

Looking at various t-tests gives

Combined values percentage A versus B two-tailed t-test gave $p = .017$.

Copier percentage A versus B two-tailed t-test gave $p = .054$.

Software percentage A versus B two-tailed t-test gave $p = .193$.

The individual tests (copier or software alone) are not significant, but the combined value is. Coupled with the finding that the differences between the two are statistically significant, this value should be probably be discounted. Thus, we are left with the somewhat unexpected result that the percentage of global comments does not change between grades.

Both A and B students made a similar percentage of global comments. Unlike the analysis of raw total comment values, which showed A students made a statistically significant higher number of comments, when viewed as a percentage, the differences for global comments are not significant. Both A and B papers received similar percentage of global comments. However, this must be viewed in light of the total comments and understood as that although the percentage difference was not significant, overall an A paper had more comments (a difference that was statistically significant).

## A Closer Look at the Global Comments

In a second round of coding, the global comments were coded based on what they discussed. Each comment was placed into only one category, although many could have been dual-coded. The four categories were as follows.

- Graphics. Comments about adding or modifying a graphic. (Comments to move a graphic to a different section were coded as organization.)
- Organization. Comments about moving, deleting, or significantly reworking a section.
- Audience. Comments that specifically address appropriateness of audience issues.
- Details. Comments that gave guidance on how to improve the text. Since these were global comments, the guidance was high level such as "why does this section not include specifics, as do the other sections?"

Table 6.3 gives a breakdown of the coding results. The rather extreme difference in number of comments in each category for A and B papers results from both B papers having fewer total comments and there being fewer B papers (see Table 6.1). B students suggested moving material, but didn't get into details of how to improve the text.

Because of the highly unbalanced number of comments between A and B papers and the low number of B comments, a detailed statistical analysis would not be appropriate. Nor would it be very informative. However, a deeper look at the comments themselves—done in the discussion section—does prove

**TABLE 6.3** Category breakdown of global comments.

| | Grade | Graphics | | Organization | | Audience | | Details | |
|---|---|---|---|---|---|---|---|---|---|
| | | count | % | count | % | count | % | count | % |
| Total | | 57 | 27% | 56 | 26% | 28 | 13% | 69 | 32% |
| Copier | A | 23 | 27% | 19 | 23% | 14 | 17% | 27 | 32% |
| | B | 4 | 22% | 9 | 50% | 1 | 6% | 3 | 17% |
| Software | A | 21 | 26% | 21 | 26% | 12 | 15% | 27 | 34% |
| | B | 9 | 28% | 7 | 22% | 1 | 3% | 12 | 38% |

informative about how students make comments and where pedagogical instruction could change to improve them.

Note the low ratio of comments on both organization and audience between the A and B students.

## Discussion

In this study, too many of the students—and remember these were graduate students—performed a line edit of the text as written without evaluating if each sentence, paragraph, and section made sense within the overall document. The low number of audience- and detail-level comments, especially from the B students, shown in Table 6.3 highlights this problem. Instead of editing to ensure the text communicated, they took the text as provided and did a sentence-level edit. Whether or not that text made sense was not considered. I have, unfortunately, talked to too many practitioners who have that same view, with comments such as "Our engineers send me what they write, and I make it grammatically perfect and format it." Whether or not that text makes sense or communicates anything is the responsibility of the engineer (author) and defined as outside the scope of the editor's duties. But ensuring the text communicates should be the first and foremost editor duty.

Of course, editing comments will never all be global. Any text contains both sentence- and paragraph-level issues that need to be addressed along with global issues. Even with the wide variation in the number and quality of global comments, some factors are clear:

- Papers with higher grades had more global comments and a higher percentage—but not statistically significant—of global comments.
- Comments showed a lack of consistency in identifying global issues.
- Lower-performing student editors had difficulty identifying any global issues.

Higher-grade papers contained both more global comments and a higher percentage of global comments. Of course, as a pedagogical issue, simply stressing making global comments will probably not improve student grades or, more importantly, their ability to edit. The underlying factor is that the A-level comments took the overall paper structure into consideration; in other words, the editor looked at and commented on the paper from a global perspective. Techniques of teaching how to perform that view still need research. Existing technical communication research focuses on taking that global view when writing (as an author); we need to figure out how to translate and extend that research to editors. Part of the problem may be that the students have not fully conceptualized how to create a well-organized paper focused on communicating with an audience—a writing problem. They may be able to do it, but can't explain how they do it. Thus, when faced with editing, they can only see "here's what I would do" but can't explain how an author can revise their text and can't put the comments at a specific enough level.

Table 6.3 highlights the problem of the B students seeing global issues, especially the audience-level problems. They seemed relatively ok with suggesting more graphics and major organizational issues (most of the nine organizational comments were "move this"), but not with the "does this text communicate the information?" level of issues. The instructions for both assignments included the line "In your edit, recommend headings, lists, graphics, and tables to improve the overall look and readability of this report." Thus, they should have been primed to look for places to change or insert graphics. Interestingly, none of the comments mentioned breaking up a paragraph to contain a list.

Next, we take a closer look at the comments themselves to see how the global comments reflect issues of editing for human–information interaction. Both of the papers had a few major global issues that I expected all editors would comment on. However, there was a lack of consistency of editors seeing higher-level issues within the text. Although not explicitly coded for this study, the paragraph-level comments also suffered that same inconsistency.

Reviewing the total comments for what I considered the top three problems highlighted the lack of consistency. None of these three major problems were noted by more than three to four editors and were completely missed by many. I picked the three examples discussed below because they all show different aspects of editing for human–information interaction. All three affect how the information is perceived and evaluated, that is, the fundamental human–information interaction issues authors and editors must consider.

- The copier paper had an early bullet list that gave the requirements for the new copier. One of those requirements was "Maximum size of 46" wide and 26" deep." The size of the copiers was never mentioned during the evaluation.

  o Early parts of a text promise to provide specific information. Does that information get provided? And, more importantly, does it provide it in an effective manner?

- A formula for calculating the cost per copy was given excessive explanation and was formatted in Courier. This is a basic formula that managers would know and then, after being explained in too much detail, it was not used to estimate the cost per copy for each of the copiers.

  o Every piece of information should be questioned. Why is this information included in the text and is it in the proper place? For an editor, answering these two questions clearly can be a difficult task. In this example, the formula may be new for the editor, but the text is not for an editor, it is for a different audience. A mantra I constantly hear within the usability community is "you are not the audience." Editors need to learn this mantra too, and evaluate all of the text's information with respect to it. In this chapter I'm using the phrase human–information interaction, but perhaps it should be changed to *audience–information interaction.*

- The copier paper conclusion was written as a single long paragraph that summarized the review of all three copiers in a jumbled, inconsistent manner. It needs to be rewritten into multiple paragraphs with lists/tables.

  o How effectively the information *communicates* needs to be examined. Especially from student writers, I see a lack of willingness to use lists, probably because of many years of only writing long dense double-spaced paragraphs. A data dump of everything relevant onto the page is considered sufficient; the information is there, end of story. Thinking of how well a piece of information communicates is a new concept. One issue is a "forest and trees" problem. All of the information exists within the paragraph. But the random tree pattern needs to be reworked into a well-sculpted forest.

Finally, four papers were not coded because they failed to conform to the assignment requirements and, in addition, the B papers were generally lacking in overall comments, including global comments. Taken together, these are too high a percentage of the class. At some level, it seems that these students were failed by their undergraduate programs because they never learned how to look at texts as multilevel products and how to analyze overall text organization. At some level, we, as the instructors, must blame ourselves for this attitude. From their freshman composition days, students are told to "edit your paper before turning it in." In other words, they learn that editing equals proofreading. I've heard many late-in-the-semester discussion comments where the student thought that different aspects of a word-for-word copyedit were what they would spend the semester learning. They expressed surprised at how broad editing really was.

## Classroom Implications

Based on the results, there are some implications for teaching editing for human–information interaction and global-level editing in general.

Editing for human–information interaction and effective communication puts the text's overall content at the top of the editing priority hierarchy. I doubt if many will dispute this comment, but moving it into editing practice with pen in hand (real or metaphorically) is not as easy. At a very real level, one of the issues we need to be teaching is not just "how to edit" (whatever that phrase means) but teaching editing as a method of analyzing a text to ensure it is communicating its message effectively and efficiently.

A simple lecture will not achieve getting a student to implement the difference in their editing. They may be able to explain the difference, but will probably not reflect it in their editing. Focusing on proper comma usage is easy. Making comments to improve a sentence is easy. Making comments on how the overall document works is not so easy. At the cognitive level, dealing with document-level communication issues imposes a high cognitive load, especially for people learning to edit. Unfortunately, the end result is often sloughing off the load rather than dealing with it. Load sloughing is how people deal with cognitive overload. Cognitively, people cannot work in an overload state; one way or another task load is reduced (Thuring, Hannemann, & Haake, 1995).

Editing for human–information interaction means the editor must consider how people interact with the information. Ok, that's a rather duh statement. But how does an editor do that? The fundamental pedagogical problem is how to teach students to do an information interaction analysis of a technical document; how to read and evaluate a text before making comments. How well does the document communicate? The analysis must evaluate both what aspects are supporting information interaction and what aspects impede the content from communicating its message. An editing course is probably the first time students are forced to analyze existing texts rather than produce their own. The editorial analysis focuses on details and shifts from a high-level "look at this text" to "is this text organized effectively?" and "is this text what the different audiences require . . . too much or too little content?" An analysis must ask those questions at the document, the section, and the paragraph levels—in that order.

At a copyedit level, the focus on "edit the text as provided" is expected and is more or less the definition of copyedit. But editing for human–information interaction occurs as part of the developmental edit and requires a clear understanding of the document's audience. Students are constantly told to have an audience in mind when they write, but the reality is that they only write for the instructor and "audience considerations" get subsumed into writing for the instructor. Being able to actually examine a text with respect to how well it addresses an audience is a learned skill; a skill that we should be teaching in all writing classes at all levels, but most certainly at the graduate level. Once analyzing for audience is learned, the text organization issues should be apparent and are essentially subsumed by organizing for audience needs. Perhaps a course focused on audience and audience needs should become a staple of the technical communication curriculum.

Human–information interaction looks at how an audience interacts with and uses the information. The idea of "interacts with and uses" information is different from a typical audience analysis that gives little more than demographics or sweeping generalizations based on job/task. Translating the audience analysis into a form useful for either writing or editing is not an easy task. What an audience does with the information and how to support that interaction needs to be explicitly taught. And, of course, this audience consideration gets highly complicated since any realistic text has multiple audiences, all of which must be considered and addressed.

Seeing organizational issues requires seeing the overall structure of a text. The online versus paper editing argument is old, and has long since been won by the online proponents (Dayton, 2003). However, viewing a document on a computer screen reduces the amount of visible text too much. People discount text they can't see and having to mentally maintain the content of different sections of the document imposes high cognitive workload. Teaching students to print a document and spread it out on a table could be one way to help teach high-level editing. I think this would be a great classroom-based research project (one I can't perform since my editing teaching is online): When the entire text is before them and viewed a whole, can students better identify the organizational problems?

Creating comments is an important classroom exercise. Editing courses need to teach three things about comments.

1.  How to write comments.
2.  What to comment on.
3.  How to identify what to comment on.

The first two points are standard editing course pedagogy. I'm not as sure about the third and wonder if it gets rolled into what to comment on. In other words, students are told to comment on poor organizational issues, etc.

The lack of recognition of audience and organizational problems is not just an editing issue; it's a fundamental how-to-create-content issue. It is not an editing course issue; it is an issue that spans the entire curriculum. Practicing technical communicators do not work on documents as single individuals—the writing style they learn in academia—but as part of teams with each member contributing different parts. Different parts that then need to be reshaped into a coherent final product by the technical editor. At the curricular level, we should be striving to teach this level of document creation.

Editing textbooks call for multiple passes through a document. I'm not sure how many students actually do that; or they may do multiple passes, but each pass looks at everything. I'm not sure how to effectively teach the need for multiple passes. Even the two texts in this study were not long enough to allow for focused class-based discussion of each pass. I'm imagining an activity such as "do pass 1 where you focus on X. We'll worry about other aspects in later passes." Then, after they do that pass, it is discussed, and the class then cycles through pass 2, 3, etc.

Reshaping and suggesting extensive revisions may change the author's voice—a perceived problem that does not apply to technical editing. Each time I teach editing, I wait for someone to bring up "not messing with the author's voice." They hear about the importance of not interfering with the author's voice in freshman composition and in any creative writing courses they take.

However, I point out that technical writers rarely have *a voice*. The content is not the author's voice but is the voice of the organization for which the author is writing. (If they are editing creative writing, different voice rules apply, but that is outside the scope of a technical editing course.) Every document that comes from an organization should sound as if one person wrote it. Even documents with multiple authors must sound as if one person wrote it. In other words, the job of an editor is not to preserve an author's voice, but to preserve the organization's voice and reputation. When students are too concerned with preserving voice, they fall into a mode of basic copyedit and accepting everything else. Paragraph and global-level issues get ignored.

Perhaps the editing assignments need to be more specifically focused as opposed to "do a comprehensive edit of the text." In undergrad editing classes, I've at times banned any copyediting of a text as a way to force the students to focus on higher-level issues—yes, I specifically say "do not mark comma errors, etc." Being told to only address global-level issues that prevent the document from clearly communicating forces them to skip all grammatical concerns and to think about the text at a higher level. Students need a mind shift from the structure of student academic papers to corporate-style technical documents. The sentence-level comments displayed an excess of comments about APA formatting issues or general formatting, such as "this heading needs to be 14 point." The "papers must conform to APA style" gets deeply drummed into students and then, in this course, we are asking them to move past that. It's clear that simply telling them that once or twice in class does not carry over. When they are engaged in the high cognitive load activity of editing, they fall back onto prior knowledge.

It would be interesting to see what would result if I gave the two assignments evaluated for this chapter to the class and told them they were only allowed to make structural and/or organizational comments; that copyedit and formatting issues were to be ignored. I sometimes think that some students—typically those we would define as the lower half of the class—view the assignments through a lens of just having to make enough edits or comments to get a good grade. I've never been asked this question directly, but I do suspect some students estimate the minimum amount of work they must put into the assignment based on it. If editing is viewed as a simple count, then fixing a verb tense and suggesting combining and reorganizing two sections both count as 1. Clearly, as instructors, we see those two actions as very different; we need to figure out how to get the student to see that distinction. Only then can we get them to focus on the higher-level edits that will improve a text.

## Conclusion

Comprehensive editing requires editing for high-level organizational and audience issues to answer the basic question of "does this document communicate its information efficiently and effectively?" As such, it is important that editing courses teach students how to perform this level of edit. More than just the comprehensive edit as defined in most editing textbooks, editing for ensuring a document communicates its information efficiently and effectively requires editing for human–information interaction. Looking at the editorial comments from graduate students, we found that their editing skills can be significantly improved in this regard. They understood the basics of editing, but still needed to gain skills in analyzing and commenting on documents for high-level organization issues.

Thus, when we teach developmental editing, we need to teach how to analyze documents at a high level. In fact, I'm clearly making a claim that our pedagogy of "teaching comprehensive editing" is inadequate and needs to be expanded to incorporate human–information interaction.

## Open Research Questions

Although not explicitly coded for this study, the paragraph-level comments also suffered that same issue with inconsistent identification of problems. A fundamental question that needs research here is whether the quality of the student comments reflects their ability to create a text or just to edit it. Depending on the answer, the effective pedagogy to address the shortcomings would be different. It would be an interesting study to have professional technical editors perform an edit on these two documents and examine the consistency of their comments. It would also be interesting to have students (probably in a technical writing course) do a full rewrite of these two texts and compare what they produce against what editors' comments aim to produce.

In a related idea, if a person is revising their own text, obviously they then make any changes they feel are needed. When editing another person's text, rather than making the change, the editor should create a clear and useful comment explaining the problem and potential solutions. This brings up the question of whether or not an editor's ability to perform a comprehensive edit is related to that person's ability to revise texts. How would a "revise this text" assignment compare with an "edit this text" assignment? They should produce similar results. I doubt if they would have similar results, but the difference could drive curricular changes.

But more importantly, from a practical and pedagogical view, we need to figure out how to teach students to recognize text problems. Knowing how to create a good editorial comment fails the student if they can't identify the problems on which to comment. Discussions on how to write editorial comments occur in the classroom. We discuss the proper way to address issues, such as with questions or direct statements, etc. However, a deeper issue that cannot be addressed with generic "how to write comments" discussion is not the "how to say it" but the "what to say." Not just global comments, but how to analyze the problems and determine what to comment on. Global-level comments and comments that address human–information interaction rarely point out trivial fixes. They are not at the level of reworking a sentence and moving on to another problem; instead, they risk incurring major changes or restructuring to, at a minimum, the entire section. The editor needs to explain the logic enough to convince a writer to invest the time in that rework. Sometimes I feel student editors forget that authors submit what they view as a solid finished product.

## Pedagogical Considerations

- Along with teaching just "how to edit," we also need to teach editing as a method of analyzing a text to ensure it is communicating its message effectively and efficiently.
- Teaching editing at different levels—sentence, paragraph, global—each requires an explicit pedagogical focus and specific exercises.
- Teaching editing should focus on issues of "how well does a document communicate?" The editorial analysis must evaluate both what aspects of the text are supporting information interaction and what aspects impede the content from communicating its message.

## References

Albers, M. J. (2008). Human–information interaction. *28th Annual International Conference on Computer Documentation.* Lisbon, Portugal, September 22–24.

Albers, M. J. (2012a). *Human–information interaction and technical communication: Concepts and frameworks.* Hershey, PA: IGI Global.

Albers, M. J. (2012b). Information analysis: A complex socio-technical problem. *Symposium on Communicating Complex Information.* Greenville, NC, February 24–25.

Albers, M. J., & Marsella, J. (2011). An analysis of student comments in comprehensive editing. *Technical Communication, 58*(1), 52–67.

Anderson, S. L., Campbell, C. P., Hindle, N., Price, J., & Scasny, R. (1998). Editing a web site: Extending the levels of edit. *IEEE Transactions on Professional Communication, 41*(1), 47–57. doi:10.1109/47.661630

Buehler, M. F. (2003). Situational editing: A rhetorical approach for the technical editor. *Technical Communication, 50*(4), 458–464. Reprinted from 1980.

Dayton, D. (2003). Electronic editing: A survey of practices and attitudes. *Technical Communication, 50*(2), 192–205.

Dragga, S., & Gong, G. (1989). *Editing: The design of rhetoric.* Amityville, NY: Baywood.

Eaton, A., Brewer, P. E., Portewig, T. C., Davidson, C. R., & Portewig, C. C. (2008). Examining editing in the workplace from the author's point of view: Results of an online survey. *Technical Communication, 55*(2), 111–139.

Gulliksen, J., & Lantz, A. (2003). Design versus design—From the shaping of product to the creation of user experience. *International Journal of Human–Computer Interaction, 15*(1), 5–20.

Hayhoe, G. (2002). Core competencies: The essence of technical communication. *Technical Communication, 49*(4), 397–398.

Lanier, C. R. (2004). Electronic editing and the author. *Technical Communication, 51*(4), 526–536.

Rude, C. (2006). *Technical editing.* New York, NY: Pearson.

Samson, D. (1993). *Editing technical writing.* New York, NY: Oxford University Press.

Thuring, M., Hannemann, J., & Haake, J. (1995). Hypermedia and cognition: Designing for comprehension. *Communications of the ACM, 38*(8), 57–66.

Van Buren, R., & Buehler, M. F. (1980). *The levels of edit* (2nd ed.). Pasadena, CA: Jet Propulsion Lab.

Walkowski, D. (1991). Working successfully with technical experts—From their perspective. *Technical Communication, 38*(1), 65–67.

Willen, M. (2004). Teaching effective feedback skills. *Intercom, 51*(4), 21–22.

Winsor, D. A. (1993). Owning corporate text. *Journal of Business and Technical Communication, 7*(2), 179–195. doi:10.1177%2F1050651993007002001

# 7

# CONCEPTS IN TECHNICAL EDITING TECHNOLOGIES

## What's Important in Practice?

*Clinton R. Lanier*

## Chapter Takeaways

- Technical editing technology is not constant, but differs depending on industries, organizations, and the preferences of authors and reviewers.
- Technologies share a limited number of common features, including collaborative features, version control, and validation.
- Most reviewers prefer Microsoft Word, or at least Word-like tools, and so this was the most frequently cited tool used by the study participants.
- Although most used Microsoft Word, it was often looked down on by other participants and even by those who used it because they did not consider it a "real" publishing tool.
- There are a variety of tools used for myriad purposes.
- Organizations may mandate the use of technology that is not preferred by the authors or reviewers, and so editors may have to decide what technology better fits the context.
- The technology chosen by organizations may not always be the best tool, and editors may have to find other tools to supplement it.
- Technology is becoming more specialized for different industries. It is also becoming more automated and seems to be moving to online platforms that enhance or better facilitate collaboration.

## Introduction

For decades, technical editing was primarily carried out with paper and pencil (or pen). This began to change with the introduction of desktop publishing platforms in the 1980s, but the change (as discussed in studies such as Rude & Smith, 1992, or Duffy, 1995) came slowly. However, the practice has evolved, and today relies on a number of different technologies (as well as the traditional papers, pens, and pencils). To understand and identify these technologies, this chapter discusses the results of a forum conversation with practicing technical editors. More importantly, using these interview discussions and relevant literature, the chapter will report on how current technologies are changing, and what these changes mean to those entering the profession.

The following section discusses past studies about technical editing technologies. I then present the methodology used to create a virtual focus group of technical editors, and to report the major points of the conversation held with these editors. Finally, I discuss the implications of these points, including major lessons that can be taken from them.

## A Brief Review of Technical Editing Technology

The potential for technology to better facilitate technical editing has been recognized for decades, but its actual use has been very slowly embraced (for a full history of literature regarding technical editing technology, see Dayton, 1998). Here I want to focus on important milestones in the progression of technical editing technology. This will better frame the results from the current study and inform the discussion that follows.

Though the use of technology in the editing step of technical editing was widely derided for years (Dayton, 1998), Farkas and Poltrock suggested in 1995 that "computer technology can provide impressive support for many group-writing activities" (p. 155). However optimistic they were about the use of software for technical editing though, they also recognized the drawbacks listed in previous research: "there is . . . one part of the review process in which computer support is considerably less effective: editing" (p. 155). In their study of technical editing at a large technology corporation, they found that the actual act of editing (that is, marking the mistakes or changes in the document) took place primarily via pen and paper—the most traditional method of hard copy markup and editing. This is not to say that technology was excluded from the technical editing process completely. Rather, it was often used for transmitting documents (through email, for example) for communication with reviewers and authors, and for document version control. Version control essentially uses a database that keeps track of the newest version of a file so that everyone editing or reviewing that file is working with the latest. This ensures that team members are not wasting time on old versions and that any change made affects only the newest document or file. All of these aspects are important to the editing process in total.

At the time Farkas and Poltrock conducted their study, the reasons for resisting editing via software or word processing tools were many (and quite understandable). To begin with, the tools simply were not advanced enough to record editing markup sufficiently. Changes did not stand out nor was there any way to tell who made the changes, so the only way to tell what was changed was by comparing the edited document with the original. There was also no facility for inserting comments, so questions at the point in the document where the question arose could not be asked. But, while editing itself was not taking place through software, the other items they noted (the transmission of documents, communication, and version control) were still major improvements to the technical editing process. Editors and authors, for example, collaborated at a distance through electronic communication methods, the electronic transmission of documents cut down the turnaround time in the review and editing steps, and version control ensured a more efficient process.

Interestingly, though, in a subsequent study completed in 1999, Dayton (2003) reported that the number of technical editors using software as an editing platform (dubbed "e-editing") was 46%, while 54% remained using the traditional hard copy methods (p. 46). His study focused primarily on members of the Society for Technical Communication (STC), so may not have been completely representative of the profession as a whole, but it was still the most comprehensive study at the time. In his findings, Dayton listed the primary type of technology used (11% of his respondents) as "automatic change tracking in software" (p. 46). By this point (four years after Farkas and Poltrock), Microsoft Word's track changes feature had been in use since the introduction of Microsoft Windows 95 in late 1994 (Campbell & Lawrence, 1995).

Apparently, the track changes feature had a major impact on the view and use of e-editing tools in a very short time. The track changes feature, in fact, addressed many shortcomings of editing software that editors had cited in previous studies. With Microsoft Word 95's introduction, editors and reviewers could now easily find suggested changes to the document, and then accept or reject those changes. The acceptance of this technology was still slow to come. As late as 2003 the U.S. Government laboratory I worked for was transitioning away from hard copy, paper-based technical editing (not copyediting, but comprehensive technical editing) to e-editing using Microsoft Word (Lanier, 2004). It took over a year for this transition to take place, but eventually all editing took place electronically.

Since its introduction, Microsoft Word has become one of the major platforms for collaborative writing, allowing a number of writers, reviewers, and editors to efficiently work on the same document. However, over the past ten years, other platforms have been created that mirror many of Microsoft Word's editing features. Google Docs, for example, also allows users to insert and track edits and changes to documents. It also allows comments to be inserted by multiple collaborators, and, because it is cloud-based, allows multiple users to work on a single file synchronously. Likewise, Open Office, an open-source software suite resembling Microsoft's Office software, also records edits and changes on its document platform. And even specialized publication platforms, like MadCap Flare, a specialized software used for creating and publishing structured documentation, have integrated edit tracking features as well.

The technology used for editing, in short, has clearly evolved in a way that technical editors find easier to use and more efficient. The features of that technology, optimistically prophesied by Farkas and Poltrock, has had a profound impact on the editing process. However, a current understanding about the intersections of technical editing and technology is needed, and so this chapter tries to fill that need. The next section discusses the methods I used to better understand what technologies editors are currently using, and why.

## Methodology

To better explore and understand current technologies used in technical editing, I chose to conduct online forum discussions with practitioners. To facilitate this I began a conversation thread with two groups hosted by the LinkedIn social media platform. These groups, the Society for Technical Communication and Technical Writer, are virtual meeting places for technical communicators who post questions, tips, and advice.

Online research mechanisms, such as forums and surveys, have been widely studied and are found to be efficient and trustworthy methods of gathering information from a population. Wright (2005), for example, suggests that online research tools give researchers great access to specific populations that may not be accessible otherwise. Further, online research methods, according to Granello and Wheaton (2004), have a number of benefits, including "reduced response time, lower cost, ease of data entry, flexibility of and control over format, advances in technology, recipient acceptance of the format, and the ability to obtain additional participants" (p. 388).

The Society for Technical Communication (STC) is the largest trade organization representing technical communicators in the world. It is an international organization with over 6,000 members worldwide and numerous special interest groups. The LinkedIn STC Group has over 12,000 members (you need not be a member of STC to be a member of the LinkedIn group), and is very active, with members posting to it daily.

The Technical Writer LinkedIn Group has over 14,000 members and describes itself as a resource to "'Give and Take' knowledge and create a long lasting enthusiastic group who can provide solution [sic] to various Pros and Cons pertaining to Technical Writing in all the Industries." This group, too, is very active with members creating new discussion threads at least two to three times per week.

To initiate the conversation, I created the following post.

> I'm a Technical Communication instructor at New Mexico State University and am currently working on a chapter for a new book about Technical Editing. I'm hoping any Tech Editors out there can help me answer one simple question:
>
> – What technology or tool do you use to actually perform editing?

When I worked as a Tech Editor we simply used MS Word with Track Changes, but at IBM we used something proprietary—what do you use today?

- Finally, could you share your view of the pros and cons of the technology you use?

Instead of seeking many responses through a survey instrument, my goal was to conduct a virtual focus group. Though the number of responses would be much smaller, they would be much more detailed. Further, this type of online forum allowed me to explore the answers, ask for clarification, and create conversations between and amongst the participants. I chose to use a focus group-like approach because research has shown that it is more flexible than a survey and can provide a greater amount of information when used to explore a topic (Flick, 2014; MacNealy, 1999).

Rather than simply answering a question, focus group participants use the group itself to shape the conversation in ways that clarify, specify, and stimulate further responses. Therefore, unlike one-on-one interviews, focus groups allow participants to interact with each other to elicit even more information. Focus groups have commonly been used in marketing and usability research, but have also been used in exploratory research in technical communication to establish considerations in new fields (see, for example, Farkas, 1987). The primary goal in using a focus group is to reveal issues for further research or to further explore known issues through the content of the discussions.

There may be concerns about using an open-ended question for this study. A chief concern may be that respondents have too little information to form a proper response. Further, some may consider open-ended questions much more difficult to analyze than closed-ended questions (Roberts et al., 2014). And, after all, any interpretation of the responses could be drawn from the theoretical expectations of the researcher. Such concerns, however, can be assuaged by using a structured coding process (Flick, 2014).

## Participants

In all, 27 participants created 44 different forum threads. Some of these were direct responses to my initial question, while others were responses to other participants or to me when I asked further questions about their initial response. Of the 27 total participants, there were 15 men and 12 women. The average number of years of experience in technical communication was 14.92 years, with a median of 16. The countries represented by the participants were the United States (18), Canada (4), England (2), Australia (1), India (1), and Ireland (1). The industries represented included technology, software, finance, science, education, and training.

## Coding and Analysis

The resulting responses were compiled into a spreadsheet for analysis. Using Flick's (2014) method of coding, I began analyzing the responses and creating categories within which they could be grouped. Flick utilizes the stages of open, axial, and selective coding. In the first step (open coding), the researcher begins tying together distantly related concepts into large, loose categories. These are subsequently refined in the second step (axial coding), as relationships begin to emerge between different, smaller categories. Finally, the last step (selective coding) finds connections between different subcategories and generally attempts to find causal or other types of relationships.

In the open coding stage of my analysis, I began selecting words, phrases or ideas that were common across numerous responses. For example, I quickly found that Microsoft Word was recorded in many of the participants' comments, so I created a column labeled "MS Word" and placed each occurrence into it. If the response included context (such as how the participant felt about MS Word), then I included that as well.

During the axial coding stage, I considered small differences between comments and either left them in separate columns or placed them together. In the example from the previous paragraph, I could have potentially placed a response discussing "edit tracking" into the same category as MS Word, because this is a feature of Microsoft Word. However, in such a case I made a separate subcategory for "edit tracking" and began placing anything discussing edit tracking into it.

During selective coding, the last stage, I tried to find relationships between or among the categories. If relationships existed I placed the columns next to each other. In the end, four main relationships were identified and thus four categories emerged to contain the subcategories created in the axial coding stage. The process does not end by creating these four categories. In fact, these are only convenient ways of placing subcategories that have a loose but certain relationship. Once all the categories have been formed, then the analysis continues by examining each comment individually and looking at large-scale implications of the information.

## Results

In brief, the following software tools were discussed as being used by the 27 participants in the group (Table 7.1).

The discussion responses about these technologies focused on four general topics, which I used to create categories for organization (Table 7.2). Following the table is a discussion about the responses each category contained.

## Convenience

Out of the 27 participants, 19 of them were adamant that the technology used is not necessarily based on what is most efficient or effective, rather it is mainly

TABLE 7.1 Technologies used by participants.

| Technology used | Number of participants who use technology |
|---|---|
| Microsoft Word | 7 |
| Visio | 6 |
| Adobe Acrobat | 3 |
| Camtasia | 3 |
| Google Docs | 5 |
| LucidChart | 3 |
| MadCap Flare | 3 |
| FrameMaker | 2 |
| InDesign | 2 |
| PowerPoint | 2 |
| Captivate | 1 |
| Illustrator | 1 |
| NaturalReader | 1 |
| Overleaf | 1 |
| PerfectIt | 1 |
| Photoshop | 1 |
| Snag-It | 1 |
| StyleWriter | 1 |
| TechScribe | 1 |

TABLE 7.2 Categories derived from forum discussions.

| Categories |
|---|
| Convenience |
| Version control |
| Collaboration |
| Validation |

based on convenience for the author. The fourth response I received noted "I use a variety of methods, depending on the author and the software he or she has used." This theme was repeated many times, but was especially present when discussing specialized writing or publishing tools.

For example, while 15 participants said that texts or documentation were published on specialized platforms, reviews took place with more common software, because, as one noted, "most would-be reviewers, especially for some reason technical reviewers such as software developers, are averse to download-ing, installing, and learning a special-purpose reviewing tool." This was repeated by a second participant, who said that "SMEs and other reviewers have no need for such tools any more than most tech writers will require an Integrated Development Environment (IDE) of the sort used by software developers to collaboratively develop complex applications." A third said similarly, "most reviewers and SMEs, especially technical ones, find reviewing documentation onerous enough. They tend to drag their feet if an additional layer of process or complexity is imposed upon the process."

Of the tools most often cited as used because of convenience, Microsoft Word (MS Word) and Google Docs were both mentioned prominently (seven and five times, respectively). For example, one participant noted she used Microsoft Word, "because it was so ubiquitous among stakeholders . . . it is accessible to all of my clients." Another suggested, "even technical professionals are averse to change at times, and Word has the advantage of familiarity." Still another argued,

> For better or worse MS Word nearly always ends up being the path of least resistance. Pretty much everyone has a copy of it and can fumble their way through the use of the program's Track Changes feature. Even my attempts to send reader-commentable PDFs are often met with 'can't you just send me a Word file?'

It was this "path of least resistance" that another noted when she stated that they publish documents using Madcap Flare, however, "getting SME reviews with it is not as easy as a Google doc/MS Word with track changes."

And even when one tried to move reviewers and authors to the specialized publishing platforms, he ran into resistance that caused problems: "eventually someone needed to get something done fast. (Looming deadlines will do that to ya'.) So a sort of black market in Word files containing 'oh, one last thing' add / move / change requests came into existence."

Still two other participants (both freelance editors) said that the use of technology was based on the environment of the employer. One simply said, "I use what employers provide," and the second said that, "you have to be flexible and ready to use whatever they throw at you."

## Version Control

One of the first points made about the technology used was not so much about the writing or editing technology, but about the use of SharePoint, a team collaboration software made by Microsoft. SharePoint allows multiple team members working on a document to check the document (or set of files) in and out, controlling the version and ensuring that updates are made to the most current version. Further, it keeps copies of previous versions as well. Both of these features were pointed out as important and echoed by a different participant.

Another participant noted that version control is a feature of Google Docs, which she used in part because of its versioning feature: "It tracks changes and you can easily revert to old versions—plus it ensures people are always working on the most current version (that was an issue when using Word)." Google Docs was used by an additional four other participants, and in all of their responses versioning was named as a reason.

Still another software tool that facilitates versioning is MadCap Flare. One participant told the group "more recently I've been using a version control system supported by Flare for tracking changes." Flare was an oft-mentioned technology by the group, but this was the only participant who pointed out its version control abilities, or suggested he used it because of that ability.

A participant, whose employer publishes LaTeX documents, said that he used Overleaf, an online collaborative LaTeX writing and production platform, in part because it controls document versions. LaTeX is a "high-quality typesetting system" popular in the science disciplines (www.latex-project.org). The participant further suggested that they had tried other solutions as well before coming to Overleaf, including GitHub, which he said was awful because "it was horrid for document control." He expanded on this saying that "someone somewhere would change a document, then GitHub would refuse to merge my changes to the document because of the other person's changes." In all, version control was brought up by 14 different participants at some point in the discussion.

## Collaboration

Technology that facilitates collaboration was also mentioned by 12 of the participants. Here I define *collaboration* as multiple people working together on a single text. One noted technology to assist this effort was a change tracking ability. One respondent said that this, as well as a commenting mechanism, were the reasons for using Google Docs: "More often now the software is Google Docs, which automatically tracks changes and allows for questions and answers as well—very helpful."

A different method for facilitating collaboration exists in the ability to allow multiple people to work on the document concurrently. One participant, whose organization could not use Google Docs for security reasons, stated, "I've been slowly pushing my team to use Word Online (Office 365) for collaboration because it allows concurrent editing much like Google Docs. The collaborative advantages of both versions of Word are substantial."

Without elaboration, another participant simply said she used "Adobe PDF review server. In the past I have used browser based bespoke collaborative reviewing tools." So, while not allowing concurrent editing or writing, this technology did facilitate multiple reviewers including comments synchronously.

Still another, who listed many different technologies used, suggested ultimately that what is important about the technology used is its ability to allow people to work together in real time. When discussing what tools technical editors should use, he suggested that "I'd also look for tools that provide reviewers and correctors a way to interact while stuff is being written, and tools that provide social collaboration and feedback after the fact, when the *stuff* is already online."

## *Validation*

Eight of the respondents pointed out the importance of validation tools. These are tools that match a text against some set of defined rules. The rules could be as simple as grammar or as complex as layout and format. One participant who works in the marine industry stated:

> [m]uch of my editing work is related to ASD Simplified Technical English (www.asd-ste100.org) and to terminology management. I use the TechScribe term checker for ASD-STE00. . . . The checker is primarily for ASD-STE100, but it is also good for terminology only. Many rules are applicable to plain English.

Regex—Regular Expressions—was also pointed out as a method for check-ing documents. Regex is most commonly seen in the "find/replace" feature in most word processing software programs, including Microsoft Word, Notepad, TextEdit, and Google Docs, but is also used in more complex soft-ware, like those made for writing code. One participant suggested "a 'tool' that can streamline editing, particularly massive content, is mastery of Regular Expressions (regex)." The post that followed and was written by a different participant affirmed the use of regex, and informed the group that "it may at first glance seem daunting. But some of REGEX's simplest statements provide disproportionate find/replace power. However, even though I regularly use REGEX I don't always get the syntax right the first time."

A third participant simply told us that "nothing goes out of our office with-out it going through StyleWriter—the plain English editing software." So, while they may be of different forms, it seemed that some type of validation tool was deemed necessary before a document is considered "complete."

Ultimately, the results show that there are many considerations to make that determine what tools will be used, and for what purpose. The following section more comprehensively discusses these results in context and creates a series of technology considerations for technical editing.

## Discussion

This focus group discussion demonstrated that understanding the technol-ogy used for technical editing is complex. To begin with, there is no single tool used throughout the discipline. Instead, the technology used is a response to two factors: industry and convenience. For example, the participant working in the science industry for a genetics testing company was the single person in the focus group using LaTeX. LaTeX is a "document preparation system" that uses plain text to draft content that is later formatted through markup. The soft-ware is especially popular in scientific disciplines (Gaudeul, 2007), which would explain why a genetics testing company might be using it. This was confirmed when I asked the participant why he used it and he suggested it was because of the manner in which the tool formats and displays chemical and mathematical formulas. Thus, his particular industry requires he uses LaTeX.

Similarly, the participant in the marine industry had to edit documents written to ASD-STE100 Simplified English requirements. ASD-STE100 Simplified English is a controlled language specification originally created for the aerospace industry in 1988. The specification was created to ensure that aircraft technical documentation can be translated accurately, and its rules exemplify this purpose. For example, it only uses 985 words and excludes certain auxiliary verbs such as *might, should,* and *may* because they do not exist in other languages (Smart, 2006). All aircraft manufacturers use this specification, as do many other large, international technology industries, such as the defense and maritime industries. Thus, this participant must use tools that conform to ASD-STE100 (such as the TechScribe software he said he uses).

Editors also use technologies convenient to either the employer, the author, or both. One of the most often-noted technologies used, for example, was Microsoft Word. All of the responses from those who used it indicated that their reason for using it was because it was *ubiquitous* (a word used by multiple people). The participants said that it was preferred by authors and subject matter experts because it was both common (easily available) and familiar (easy to use). And even when the publishing platform was something more specialized, like MadCap Flare for example, editors converted text to Microsoft Word to send to authors or reviewers who would provide feedback, which was then converted back to the specialized platform. Thus, the choice of using Microsoft Word was due to a matter of convenience for the author rather than an attempt to find the most effective or efficient tool. And, in fact, when a participant suggested the use of Microsoft Word, it was often followed by some sort of apology for using it.

And, to add to this point, when a respondent noted their use of Microsoft Word, they were sometimes met with slight criticism from others. One particularly vocal participant suggested that he was unsure "if Word should be taken too seriously. . . . It is a 'word processor', but mostly useless for professional layout or editing." Another used the following analogy in his criticism:

> Word is like the guest who never leaves, always pouring himself drinks and wandering about the place getting into things it shouldn't. He makes his own rules, breaks things, tells bad jokes, and ignores his more illustrious and perhaps snobbish peers.

In the end though, even this critic acknowledged its widespread use made it difficult to move away from it completely.

This finding (that Microsoft Word is used most often) is affirmed in Kreth and Bowen's 2017 study of 235 technical editors. In their study, they found that the software program most often used by their respondents was Microsoft Word, which was reportedly used by 84.6% of respondents. Their study also found that, aside from Microsoft Word, there were no other specialized software tools that were used by the majority of technical editors. Instead, their survey participants used over 40 different software programs and tools to carry out editing in their respective work places.

Though Kreth and Bowen did not further investigate why the technology land-scape was so diverse, the choice in technology use of my participants had much to do with what their employer used. There were several responses that indicated that technical editors would prefer to use something else, but they were restricted because of the preferences of the organization they worked for. One said that she was "pushing my team to use Word Online (Office 365) for collaboration because it allows concurrent editing much like Google Docs." I then asked for clarification, and specifically asked why she did not just use Google Docs. She then stated that "Google apps are blocked in our environment for security reasons. So our options are Word (desktop) or Word Online." She is making do, in other words, with what she has to work with, even though there may be better options out there.

Another editor noted the many software tools he used at his organization, and then said that he would "also like to get into (MadCap) Flare but I'm trying to find a legal long-term way to pilot it inexpensively. In my experience, when an employer finds you can do something effectively they want it." In other words, this was a tool (an expensive tool at that) he recognized as helpful to their pro-ject, but that they were not currently using. His solution was to try to find a way around his employer's restrictions to show them the value of this tool in order to eventually introduce it.

Sometimes the two parties interested in convenience—the authors/reviewers and the organization/employer—will be at odds with each other. Take this lengthy quote from one of the respondents.

> I've worked in organizations where it was decreed that content reviews would be performed with one or another tool designed specifically for that purpose. That tool would be used for mainstream reviews, et cetera. But eventually someone needed to get something done fast. (Looming dead-lines will do that to ya'.) So a sort of black market in Word files containing "oh, one last thing" add/move/change requests came into existence.
>
> Point being that most reviewers and SMEs, especially technical ones, find reviewing documentation onerous enough. They tend to drag their feet if an additional layer of process or complexity is imposed upon the process. Oops—when an official and an unofficial processe [sic] for reviews and contributions comes into being, that creates its own set of problems.

Thus, the organization decided that the process (in this case the review pro-cess) would use one thing, but actually getting the task done depended on the preferences of the authors and reviewers instead. The differing priorities could lead to many problems, such as different and wrong versions, and mistakes in editing and publishing.

Perhaps the most helpful view comes from the respondent who advised,

> There's no single method. Every company wants a magical purple squirrel employee that fits all needs and knows all software. That's not happening, but it's how HR and recruiters are hiring. So you have to be flexible and ready to use whatever they throw at you. The bigger thing is to have the brain ready to do the work, the speed with quick turnaround time, and the technical aptitude to use whatever tool they end up hiring you to use.

Though there was no single technology discussed, but instead a variety of different technologies depending on context and preferences, there were some features of the technology each had in common. Editors were quick to point out that thought should be given to version control for any technology considered. One participant discussed the use of a system without version control (GitHub) and the many issues they had with it—indicating this feature was not considered when the organization decided to use such a technology.

Another issue to be alert for with regards to versioning is the use of "other" file types or tools used. In the discussion about author preferences above, it was noted that an organization mandated the use of specialized review software, but authors found it more convenient (and more realistic) to simply use Microsoft Word. Such an arrangement, however, plays havoc with any version control system, because it is outside that system. Thus, ensuring that the latest changes have been made to the newest version of a document comes down to remembering that a change has been made outside of that newest version, and then updating it so the changes are reflected.

This was also true for some who discussed emailing files for review. The participant who wanted to use Google Docs but could not due to security restrictions said that

> Most [reviewers or authors] are still using Word and e-mailing files back and forth. [I am] Trying to push them to use Office 365/Word Online for concurrent editing, better version control, not having 70000 versions of the same doc floating around, etc.

It is easy to see, then, how quickly version control can be lost when trying to make the technology convenient for the authors and reviewers; however, it is a very important feature to remember and keep in place if possible.

Another important shared feature of the technology used was the ability to facilitate collaboration between editors, authors, and reviewers. This was most often discussed by variations of a feature allowing changes to a document to be tracked. Microsoft Word and Google Docs were both singled out as having such a feature, but also mentioned were PDF documents and MadCap Flare. These changes go hand in hand with version control, as they allow the document itself to contain the suggested changes to that document. Without this feature the suggestions have no context and it is difficult to understand where the change should go or why (Lanier, 2004). In fact, one participant lamented using email as a system for suggesting changes: "unfortunately, we also used email (Outlook) to group-edit short-form content from time to time." Imagine how easy it would be in such a system to overlook necessary or even mundane changes to the document.

Beyond tracking changes in the document, another collaborative feature discussed was concurrent or synchronous editing/writing/reviewing. In such a feature, multiple people can not only be viewing the file, but also working on it (changing it or inserting comments) at the same time. Google Docs has this ability—as do some more specialized software tools (such as Adobe PDF Review Server). Microsoft Word also includes this feature when used online with Office 365. A benefit of this feature is that real-time collaboration can take place virtually and at a distance. Team members can explain necessary changes at the time they are suggesting the change. Thus, even more context is created and the process should be more efficient.

A last important feature, though not mentioned as often as the previous two, was some method or tool for validating or checking the document. This could be as easy as a regex (regular expressions), or something more specialized, like an ASD-STE100 validator. In the case of regex, the feature is helpful to ensure that technical terms, styles, acronyms, and other elements are consistent throughout a document. When using regex it is important to have a written style guide for assistance. Another software mentioned, PerfectIt, checks documents against an internal style guide that is created in-house and installed into the software. It essentially completes the same tasks as regex, but the process is automated (versus manual find/replace tasks) and becomes a step in the workflow. A different but similar software, StyleWriter, was also mentioned. Finally, more specialized validators, like TechScribe, were also used by participants. These validators are specific to industries and to the software tools used.

The answers given in this focus group also provide a glimpse of changes in technology and how they affect technical editing. For example, though seven had specifically mentioned using Microsoft Word, there were also five that were using Google Docs, and one that wanted to use Google Docs but could not because of security restrictions (and her hope was to at least move to Microsoft Word online because it would be like Google Docs). This is significant in that it is a shift to online or cloud-based tools. The shift provides the ability for synchronous work on a file. Editors, authors, SMEs, and reviewers can all collaborate on a document at the same time. This is a far cry from the pen and paper method of editing that was so common even 15 years ago. That model was much more time consuming and often the edits made were out of context and took explanation through margin comments or even phone calls and conversations. The change to synchronous editing systems also allows teams to be dispersed geographically while still allowing them to work closely together.

Still another change is automation. Regex was mentioned by one participant as a powerful tool to check and validate documents being worked on. However, the remainder of participants who discussed validation mentioned automatic checkers. These systems are much more accurate and more efficient because they take less time to perform the work of validation. Such automation is also in place in the writing systems documents are composed in. Darwin Information Typing Architecture (DITA), for example, ensures structured content follows the rules defined within the organization in which they are composed. Software tools like TechScribe ensure that texts written for ASD-STE100 documentation follow the specifications required.

This brings up another change—the shift to specialized content structure. As already discussed, the aeronautics industry uses a particular set of required specifications, and these requirements were also moving to other industries (like the marine industry, for example). But a specialization can also be noted in other places. For example, DITA, which allows for content definition and reuse throughout documents, is commonly found in the software industry, or in any organization that mandates structured content. And one of the participants further pointed out the use of LaTeX in the science fields. This is all in contrast to documents printed out and edited manually. Now, editing in certain environments requires certain specialized software tools.

## Pedagogical Applications

Perhaps most obvious from this conversation is that there is not a single technology used for technical editing. Instead, technologies will vary according to industry demands, requirements, and author and organizational preferences. Do be aware, however, of the common, necessary features among any type of technology that will be used. Technical editors can be confident, for example, that some sort of tool will be required to keep track of the correct or newest version of the file to be worked on. Similarly, tools will be collaborative and allow authors and reviewers to work synchronously in real-time with editors and writers. This allows for conversation and should make for a better finished product. Lastly, some sort of validation is commonly used to ensure consistency in the documents or files. The tools that support such validation will also be different because no single one is used across industries. However, new technical editors must understand what is being validated and why, that such tools exist, and that becoming familiar with their general use will become increasingly important.

Another main lesson is that new technical editors should remember to be flexible. Each organization they work for in their career could potentially each use different tools to create their finished products. Some tools may be better than others, and there may be episodes where technical editors long for the technology used in a previous organization. However, keep in mind the features that the tools must have, and then try to work around those features. For example, if the tool the organization uses does not have version control, try to find a supplement to the tool that has this feature. Along with this lesson is the understanding that authors and reviewers—important components in the technical editing process—may have preferences different from those of the editors and even the organization. There may be moments when a decision has to be made about which preference to follow, or how best to provide for each. Which preference— the organization's or the author's—must be more closely attended to? This is a choice that may have to be made in context at the time, but understanding that it may come up is an important first step in making it.

If there *is* one technology that new technical editors should be familiar with, it seems to be Microsoft Word. Though derided by some, including those who use it, most recognize that there is no getting away from this decades-old word processing platform. The most common reason was its wide availability and familiarity. Authors and reviewers, some said, do not want to learn the specialized publishing programs that may exist in an organization. Trying to get authors and reviewers to examine edits and changes to documents in such specialized platforms led to long delays. To avoid these, editors simply sent the documents (with changes or edits) to the authors in Microsoft Word. However, this workaround also came with its own set of considerations, such as version control and ensuring that any updates or changes made in the Microsoft Word versions were then made in the specialized software. But, in some cases, editors told me that they use Microsoft Word for the entire process, so understanding its reviewing features, along with its layout and styles features, would be beneficial.

A final takeaway for new or entering technical editors would be a familiarity with certain industry demands, practices, or technologies. For example, ASD-STE100 uses a very specific and precise written vocabulary. Straying away from this vocabulary leads to errors and will violate the requirements of the specifications. Further, concepts like DITA for structured content are also important to understand for some industries, such as the software or hardware technology industries. Though a general knowledge of grammar and technology tools will go far, new technical editors will only be able to enter certain industries with the specialized knowledge about those industries.

## Conclusion

Ultimately, there is not, nor will there ever be a single technology used in technical editing. To approach it from the point of view of "which software should students learn how to use" would be a mistake. As this study has demonstrated, the pedagogical approach should be much more about the concepts the participants discussed. Understanding how to be flexible and work with both the employer and authors based on what is convenient for them is much more important. Students will have to adapt quickly and transport the knowledge of one type of software tool to a different context, and to a different software tool as the need arises. To do this, the students need to have a good grasp of what they should expect from these different tools, such as collaborative mechanisms, features to assist in version control, and validation. The conceptual knowledge of what is needed from a software tool, versus how to use any particular tool, will take them much further.

## Pedagogical Practicalities

- Students should become proficient at using Microsoft Word beyond the review functions. They should also understand features like styles, references, and layout. In short, they should know how to create a lengthy document with Microsoft Word.
- Students should use a workflow that familiarizes them with versioning, and should additionally work in writing groups where documents are shared and created collaboratively.
- Instructors should introduce different industry-specific standards (such as ASD-STE), and associated technologies for working with those standards so that students are aware that conventions may differ from industry to industry.
- Instructors should integrate tools that teach students to validate their information in some way. This can be accomplished with something as easy as a style guide, or with software like StyleWriter.

## References

Campbell, M. V., & Lawrence, G. (1995). *Microsoft Word for Windows 95: The complete reference*. New York, NY: McGraw-Hill.

Dayton, D. (1998). Technical editing online: The quest for transparent technology. *Journal of Technical Writing and Communication, 28*(1), 3–38. doi:10.2190/5EM1-R1TN-MMN3-3Y6M

Dayton, D. (2003). Electronic editing: A survey of practices and attitudes. *Technical Communication, 50*(2), 192–205.

Duffy, T. M. (1995). Designing tools to aid technical editors: A needs analysis. *Technical Communication, 42*(2), 262–277.

Farkas, D. K. (1987). Online editing and document review. *Technical Communication, 34*(3), 180–183.

Farkas, D. K., & Poltrock, S. E. (1995). Online editing, mark-up models, and the workplace lives of editors. *IEEE Transactions on Professional Communication, 38*(2), 110–117. doi:10.1109/47.387775

Flick, U. (2014). *An introduction to qualitative research*. New York, NY: Sage.

Gaudeul, A. (2007). Do open source developers respond to competition? The (La)TeX case study. *Review of Network Economics*, 6(2), 239–263. doi:10.2202/1446-9022.1119

Granello, D. H., & Wheaton, J. E. (2004). Online data collection: Strategies for research. *Journal of Counseling & Development*, 82(4), 387–393. doi:10.1002/j.1556-6678.2004.tb00325.x

Kreth, M. L., & Bowen, E. (2017). A descriptive survey of technical editors. *IEEE Transactions on Professional Communication*, 60(3), 238–255. doi:10.1109/TPC.2017.2702039

Lanier, C. R. (2004). Electronic editing and the author. *Technical Communication*, 51(4), 526–536.

MacNealy, M. S. (1999). *Strategies for empirical research in writing*. Boston, MA: Allyn and Bacon.

Roberts, M. E., Stewart, B. M., Tingley, D., Lucas, C., Leder-Luis, J., Gadarian, S. K., . . . Rand, D. G. (2014). Structural topic models for open-ended survey responses. *American Journal of Political Science*, 58(4), 1064–1082. doi:10.1111/ajps.12103

Rude, C., & Smith, E. (1992). Use of computers in technical editing. *Technical Communication*, 39(3), 334–342.

Smart, J. M. (August 12, 2006). Controlled English—Paper and Demonstration. Conference of the Association for Machine Translation in the Americas, Cambridge, MA.

Wright, K. B. (2005). Researching internet-based populations: Advantages and disadvantages of online survey research, online questionnaire authoring software packages, and web survey services. *Journal of Computer Mediated Communication*, 10(3). Retrieved from http://onlinelibrary.wiley.com/doi/10.1111/j.1083-6101.2005.tb00259.x/full

# 8

# EDITING FOR INTERNATIONAL AUDIENCES

## An Overview

*Kirk St.Amant*

## Chapter Takeaways

This chapter provides editing instructors with a review and summary of the

- International and intercultural factors shaping technical editing practices.
- Technical editing practices connected to usability and user expectations.
- Fundamentals of global English and English used in international contexts.
- Practices for editing to address the needs of different international audiences.
- Ancillary materials technical editors might create or use for international contexts.

In so doing, the chapter notes how instructors can teach this information and the application of related ideas to students as a part of a technical editing class or as an editing component in a technical communication class.

## Introduction—The Focus of This Chapter

Organizations are increasingly sending technical products to international audiences (Bartels, 2017). This situation means technical editors must often review content to be used by readers from other nations and cultures. Doing so requires an understanding of how to revise source (i.e., original) materials to meet the expectations of different groups. In some cases, this process involves revising source texts to make them more accessible to a wide range of individuals reading in that language. In others, it means revising texts to meet the needs of translators (Esselink, 2000; Yunker, 2003). Both situations focus on the same goal: Making materials more usable to a particular international audience.

For individuals teaching technical editing, the challenge is to provide students with the instruction needed to effectively work as editors in such situations. Editing instructors, however, might not be versed in the nuances of international editing practices. Or, they might not know how to address the complexities of editing for translation or for global audiences. One way to address this pedagogical situation is through a usability-based approach that extends core technical editing concepts to international contexts.

This chapter overviews such a usability-focused approach. The objective is to provide technical editing instructors with a framework for teaching editing students about international audiences. In examining this topic, the chapter first explains why technical editors need to know how to edit for greater global contexts. Next, the chapter reviews how editing for international audiences entails merging ideas of usability with approaches to technical editing. The chapter then presents a framework for helping students understand technical editing in terms of the needs of international audiences (i.e., user groups). It also covers specific areas students should focus on when considering international audiences. This structure allows editing instructors to connect international editing to more common practices students will likely encounter in an editing class.

## The Changing Nature of Society—The Dynamics

A challenge of teaching this topic is in helping technical editing students understand why it is important. Many students have never ventured outside of their home nation, and a few might not see the need (or prospect) of doing so. For these reasons, the teaching of international editing practices involves helping students realize the importance of understanding this area. To do so, instructors need to explain that we now live in and work in an increasingly interconnected global context that will affect future workplace practices.

The central factor driving technical editing for international audiences is market share. To begin, as global markets for technical products expand, the need to provide related documentation for users from other cultural and linguistic backgrounds grows (Rogers, 2017). In some cases, this situation means creating materials English speakers around the world can readily understand and use. Doing so creates access to the world's 1.5 billion English speakers—many of whom live in one of the over 50 countries where English is an official language (Lyons, 2017; Rosenberg, 2017). In other cases, it involves creating translated materials for individuals in a specific national or cultural market. Translation, for example, is essential to accessing broad consumer markets like Brazil, Russia, India, and China—nations that collectively represent almost half of the world's population and its consumers (Emerging Markets, 2018; Yunker, 2003). Translation can also be essential to allowing products to legally enter certain national markets or international market blocks. To enter markets in Canada, the United States' largest international trading partner, product documentation must be provided in both English and French (Medical Devices Regulations, 2017). And to enter the greater European Union (EU) market, product documentation must generally be translated into all of the official and working languages of the EU—some 24 as of this writing (Official Languages of the EU, 2018).

Other factors can also affect the need for translation within a national market. U.S. directives such as Executive Order 13166 (Improving Access to Services for Persons with Limited English Proficiency), for example, affect how medical information is conveyed to patients (St. Germaine-McDaniel, 2010). The Order could mean translation becomes a requirement when producing certain documentation for different audiences within the United States. (See www.justice. gov/crt/about/cor/Pubs/eolep.php for the text of this Executive Order.)

Moreover, the increase of certain groups of language speakers within different nations means translation can be central to the domestic success of certain products (Ulijn & Strother, 1995). The Spanish-speaking population of the United States, for example, is now larger than that of Spain (Romero, 2017). Translating for this audience could increase a U.S. company's domestic consumer base. Editing students, in turn, need to understand that the need to address issues of culture and language is both a global requirement and a local necessity. Students also need to understand the central role technical editors often play in addressing such issues.

Technical editors help craft the English-language texts for greater global audiences (Esselink, 2000; Kohl, 2008; Yunker, 2003). Technical editors are also often central to facilitating effective translation for global markets (Esselink, 2000; Sprung & Vourvoulias-Bush, 2000; Yunker, 2003). To do so, technical editors need to understand the international—and the local—dynamics affecting how an audience's national, cultural, and linguistic backgrounds affect their perceptions of information (Maylath, 2011). These dynamics are the central factors instructors need to focus on when teaching editing students about working in international and intercultural contexts.

## Audience, Usability, and Editing—The Perspective

Technical editing instructors have likely helped students realize that the concept of audience is core to technical editing (Corbin, Moell, & Boyd, 2002; Rude & Eaton, 2011; Winninger, 2013). Students have likely also learned that the goal of the technical editor is to revise materials to better address the expectations and needs of an audience. At some point, the instructor might also have helped students make an additional connection: editing's focus is connected to usability (Corbin et al., 2002; Rude & Eaton, 2011; Winninger, 2013). The revisions the technical editor suggests often focus on enhancing the effectiveness with which individuals use materials to perform a task (Corbin et al., 2002; Rude & Eaton, 2011; Winninger, 2013).

Accordingly, technical editing generally involves answering the question, "What kinds of edits do I need to make to help the intended audience use materials to achieve an objective?" The answer to this question is foundational to how technical editors should approach editing for international contexts. It is at this point that instructors can help students understand how editing for international audiences is connected to overall technical editing practices.

## Editing for Direct Use

The first concept to impart to editing students is that international editing practices represent a continuation of the core technical editing approach to audience as related to use and usability. To examine these connections, instructors should highlight the importance of identifying the needs and expectations of a given international audience.

For one kind of international audience, the objective is to use the content provided *as is* to achieve an objective (Esselink, 2000; Kohl, 2008; Yunker, 2003). This *direct use* situation generally involves creating a text that serves a broad, global audience using a common language to access information. The objective for the editor is to develop a single text that global audiences can understand and use to achieve an objective.

## Editing for Indirect Use

In other cases, the original—or source—content one creates must be translated into another language. Here, translation allows a specific international audience to use materials to achieve an objective. In these cases, translation is central to usability. As such, translators become a central figure in establishing usability by transferring meaning effectively from one language to another (St.Amant, 2013; Sprung & Vourvoulias-Bush, 2000; Yunker, 2003). The goal of the technical editor becomes that of revising materials to be more easily understood and effectively translated. Translators thus represent an *indirect audience*—for they are a conduit through which materials must pass before they can get to the intended users of those material (i.e., end users).

When working in international contexts, technical editors need to address such dynamics when reviewing a text. In certain instances, the technical editor could be reviewing work targeted at one kind of audience (e.g., global English speakers). In others, they could be editing for a very different group (e.g., translators). Often, they wind up editing for both to facilitate simultaneous international product release—when a product is distributed to as many international markets as possible at the same time (Morningside Translations, 2016). The technical editor can best address this situation through practices that meet the needs of both groups of users in an interconnected way. The challenge for instructors becomes teaching editing students how to navigate such contexts successfully. The next sections of this chapter present a framework for helping students engage in such practices.

## Editing For Global English—The Framework

Editing for global English involves creating English-language documents for as large an international English-reading audience as possible. The objective is to create one set of materials that is accessible to a broad international audience. Achieving this goal generally means revising texts to remove culture-specific items that can affect understanding (Esselink, 2000; Kohl, 2008; Norvet, 2017).

This approach reduces the cost and the time associated with developing different English-language versions of documents (Kohl, 2008). It can also make texts easier to translate (Johnson, 2016; Stein, 2015). Both of these factors are often key to being competitive in today's fast-moving global marketplace. Editing for global English can thus facilitate access to different international markets and contribute to simultaneous international product release.

This overall editing process generally includes applying foundational technical editing skills to remove culture-specific elements from a text. This process thus involves

- Reviewing the source (i.e., original) text produced by a speaker of one dialect of English.
- Identifying culture-specific wording in that source text.
- Providing alternatives for conveying meaning across cultures using that language.
- Identifying areas where the content creator implies certain relationships within a text.
- Adding elements to clarify such relationships or connections among elements in a text.

Ideally, this approach addresses elements that can cause problems when using English in global contexts (Esselink, 2000; Kohl, 2008; Yunker, 2003).

Once the instructor has introduced this approach, the next step is familiarizing students with its core practices. To do so, instructors should focus on three areas: wording, structures, and expressions. Each area will now be explained in more detail to help instructors teach the concepts.

## Wording—The Terms We Use and How We Use Them

Just because individuals use a common language does not mean communication will be easier. Rather, how individuals use words in a language can affect comprehension across direct and indirect audiences. Technical editors, in turn, need to identify such problematic factors and provide suggestions on how to remove, revise, or replace them. To do so, editors can focus on three concepts: the use of multiple referents, the varied uses of certain terms, and the need for specifying text.

## Multiple Referents

In global contexts, problems often arise when individuals use different terms to refer to the same item (Kohl, 2008; LinguaSoft, 2015; Rodgers, 2017). In English, for example, one can use the phrase "The *doctor* first consults with the patient, uses the *physician*'s perspective to diagnose the condition, and then suggests a course of treatment as a *primary care provider* would." In this case, the words "doctor," "physician," and "primary care provider" all refer to the same individual. For certain English speakers, this factor seems self-obvious due to prior exposure to these terms being used in such interchangeable ways. For individuals not as familiar with such usage, this situation can create problems.

If one is unfamiliar with the use of multiple terms for the same item, will that individual know they all refer to the same thing (one individual)? If not, confusion can result. Technical editors need to review English-language source texts to identify where authors use different wording to refer to the same idea and to select one—and only one—term to convey the related concept. The editor then needs to replace all alternative wording with that one term within the overall text.

## Varied Uses

In some instances, an author might use the same term to mean different things within a text (List, 2007; St.Amant, 2013). One, for example, can write "Set the set of sets on the set" to convey the idea "Place the arrangement of paired items on the television." These different uses of *set* could potentially confuse nonnative English speakers (including translators) who might question how the term is used in this sentence (List, 2007; St.Amant, 2013).

Technical editors need to review source text to identify where an author uses the same word to mean different things. Next, the editor needs to determine what the term actually conveys in each case. The editor should then establish one meaning and one meaning only for that term. From there, the editor needs to select alternative words to convey information in place of the original multi-meaning term (e.g., using *pairing* to replace the word *set* in key points in a text). Finally, the editor needs to thoroughly review the text to confirm that all terms in the edited text have only one meaning or use. Such steps can ensure comprehension and usability for a range of English-reading audiences (St.Amant, 2013).

## Specific Text

The relationships among ideas in a sentence are often complex and based upon cultural conventions for conveying ideas linguistically. Unfortunately, individuals who create texts are often unaware of such factors—particularly those for individual vs. collective identity and for casual relations. Addressing such items is a key component of editing for international audiences (The Acrolinx Team, 2015; Kohl, 2008; Yunker, 2003).

When teaching students this concept, it is important to discuss how individuals identify objects as singular or collective items. This factor is particularly acute when listing nouns in a sequence. If one reads "The new car needs a galface monitor, protract gauge, and sempres and itiom system to work," do the terms *galface monitor, protract gauge,* and *sempres and itiom system* represent one system of many parts? Or three or more different kinds of systems one must purchase for the engine to work? The answer can be ambiguous if the individual is unfamiliar with these nouns. If, however, one uses parallel structure (i.e., if one formats all items in a common/parallel way) the relationship becomes clearer. Consider the following examples:

> "The new car needs a galface monitor, protract gauge, and sempres and itiom system to work."
>
> "The new car needs a galface monitor, a protract gauge, and a sempres and itiom system to work."
>
> "The new car needs a galface monitor, a protract gauge, and a sempres and an itiom system to work."

In each case, parallel structure—or inserting the article *a* or *an* before items—quickly notes when nouns exist as independent entities or as parts of an overall unit (Kohl, 2008; LinguaSoft, 2015).

In such cases, the objective for technical editors becomes

- Identifying where in a text authors list items in such sequences.
- Determining what elements of the sequence are individual vs. collective items.
- Adding the terms needed to distinguish individual from collective items.

Technical editing students need to understand that this process provides the context needed to understand the meaning conveyed in such passages.

In other cases, a passage of text might not note the concept it wishes to reference. Rather, the reader needs to remember what occurred earlier to determine the meaning the author wishes to convey. The terms *this* and *that*, for example, are highly ambiguous when used independently in a sentence (Johnson, 2016; LinguaSoft, 2015; Writing for translation, 2018). Consider statements like "This is a problem" or "The user must take steps to correct that." Both examples raise the questions "What is the 'this' creating the problem?" and "What is the 'that' one must correct?" Without clarifying the wording, the reader must scan the preceding text to find any prior mention of such situations requiring correction. The text needed to clarify such items is missing (e.g., "This unauthorized use of funds is a problem" or "The user must take steps to correct that margin of error in testing environments") (The Acrolinx Team, 2015; Guidelines for writing global English, 2018; LinguaSoft, 2015).

To address such factors, technical editors need to review source texts to identify uses of *this* and *that*. They must then revise the related passage to clarify the ideas to which the author refers via those terms (i.e., the difference between "This is a problem" and "This lack of training on the part of employees is a problem"). The editor must then scan the prior content to determine what the author wishes to convey through such wording and revise the text to clarify the meaning. Instructors need to train students to follow this process when reviewing texts designed for international audiences.

### Pedagogical Applications

When teaching these concepts, instructors should focus on helping editing students understand how and when to use the following strategies to address the issues noted here.

### Multiple Referents

Addressing multiple referents involves editing students learning about

- Identifying uses of different terms to refer to the same idea.
- Establishing a single, common term to use for that idea.
- Revising the overall text to use only one word to mean one thing.

### Multiple Uses

Addressing multiple uses involves students learning about

- Identifying instances where the same term is used to mean different things within a text.
- Establishing a single use for that term within the text.
- Replacing (consistently) all other uses of that term with alternative wording.

## Specific Text

Addressing specific text involves students learning about

* Reviewing the text to determine if any ambiguities arise due to missing wording.
* Determining what the nature of the ambiguity is and what wording to add.
* Creating revised text that contains the wording needed to convey the intended meaning.

In teaching these topics, the instructor needs to emphasize that these processes involve building upon concepts editing students have likely already learned. The key is to help editing students understand that these international editing strategies use core technical editing skills to review texts in a different way for a different audience.

## Structures—The Way We Organize Information

In teaching editing students about global English, instructors also need to introduce the topic of "structures." The structure, or organization, of the phrases and sentences can create comprehension problems in international contexts because they can reflect cultural conventions for conveying ideas (Esselink, 2000; Kohl, 2008; Yunker, 2003). For this topic, the first objective for the instructor is to teach students to identify problematic structures. Next, the instructor can teach students strategies for revising such problematic structures to create an English-language text that reaches the widest possible global audience. The students should focus on three factors people use to convey ideas: passive voice, noun strings, and complex sentences.

## Passive Voice

To begin a discussion of structures, editing instructors can start with a subject they likely encountered previously: the passive voice. The standard sentence format in English is subject, verb, object, or

* Actor (the individual or entity performing an activity).
* Action (the activity the individual performs).
* Acted upon (what the actor uses to perform the action) (Frischknecht, 2015; LinguaSoft, 2015).

In the sentence "The boy threw the ball," for example, "boy" (first item in the sentence) is the subject (actor) who performs the action, "threw" is the verb/action the actor performs, and "ball" is the object or item used in performing the action (i.e., what the boy threw). We often refer to this structure as "active voice," and it identifies who is performing an action (i.e., subject/actor). A key point to emphasize to editing students is why, from an international perspective, this structure is so significant. The active voice structure is what most individuals encounter when learning English, and it often becomes the default structure for how to organize information in English.

From this perspective, the structure of passive voice sentences can create problems for nonnative users of English. Passive voice structures invert the actor–action–acted upon format by placing the object/acted upon at the start of the sentence. For example, the sentence "The ball was thrown by the boy" restructures the active voice sentence by placing the acted upon ("ball") at the start of the sentence (the conventional position of actor/subject). The passive voice structure also moves the actual actor/subject to the end of the sentence (the conventional place of the object/acted upon). This inverted structure can cause confusion for nonnative speakers who might not be familiar with it (The passive voice, n.d.; Passive voice and active voice, 2001; Rogers, 2018). This lack of familiarity could lead to questions like "Did the ball do the throwing? What role did the boy play in this situation?"

Passive voice structures sometimes complicate this situation further by omitting the actor/subject from the sentence. For example, one can say "The ball was thrown." In this case, the sentence does not directly state the actor/subject performing the related action (The passive voice, n.d.; Passive voice and active voice, 2001; Rogers, 2018). Rather, it is left to the reader to intuit who this actor is based on contextual factors (e.g., information that appears earlier in the text). Such factors can confuse nonnative English readers not familiar with this convention (The passive voice, n.d.; Passive voice and active voice, 2001).

In teaching this topic, editing instructors need to emphasize two interconnected actions to students. First, technical editors need to identify passages written in passive voice—particularly where the subject is omitted. Next, they need to revise/restructure these sentences to present information in active voice format to clarify roles and activities. Doing so removes elements that might cause confusion in international contexts.

## Noun Strings

Authors can sometimes combine words—particularly nouns—into long sequences to describe a concept. Such structures, often called "noun strings," can cause confusion. Consider the sentence "A light flux particle beam can locate certain particles found in these contexts." In this example, "light flux particle beam" is the subject of the sentence—or the actor that locates "certain particles." But what actually is that subject/actor? Is it a "light?" A "beam?" Particles? A beam of particles? Students need to understand that the task of the technical editor in these cases involves

- Reviewing a text to determine if such structures are present.
- Determining what the actual subject (vs. descriptor) of the related passage is.
- Using alternative wording to note what factors are descriptors and what the subject/actor is.

Such ideas are likely topics students have previously encountered as part of effective technical editing (e.g., creating clear content). In relation to international editing contexts, however, the instructor needs to emphasize the importance of addressing such factors to help nonnative English readers comprehend a text (Frischknecht, 2015; Kohl, 2008).

In the case of "A light flux particle beam can locate certain particles found in these contexts," the editor needs to determine that the subject is "beam." The editor then needs to establish that this term is the noun/item performing the action (i.e., continue to refer to it as "beam"). Next, the editor needs to revise this phrasing so other elements clearly describe the noun "beam" (Frischknecht, 2015; Kohl, 2008).

In this example, the "beam" is made out of "particles"—specifically particles of light in a state of flux(uation). So, a revision could be "A beam comprised of fluctuating particles of light can locate particles found in these contexts." While awkward, this structure delineates the subject/noun from the related description. In so doing, it provides clarification that can assist with comprehension for readers who have different levels of experience reading English. When covering this topic, the instructor needs to emphasize how important this factor can be when using English in international contexts.

## Complex Sentences

While simple active voice sentences are a foundational structure for presenting ideas in English, they are not the only approach available. Rather, authors can combine sentences into longer, compound sentences to convey ideas. They can also add phrases to a sentence in order to provide more detail on the dynamics presented in that sentence.

While editing students likely have seen such concepts, the instructor needs to emphasize that they can be important in international communication situations. The issue for international users is determining if the complexity of the sentence affect its usability (Frischknecht, 2015; Johnson, 2016; Kohl, 2018; Rodgers, 2017). Theoretically, the longer and more complex the sentence, the more difficult it can be for users/readers to determine the ideas conveyed in it (Frischknecht, 2015; Kaney, 2018; Johnson, 2016; Rodgers, 2017). Consider the following example:

> In order to fully initiate the overall disinfecting process, the user must start the cleaning sequence by pressing the <START> button, and once a green light indicating "on" illuminates, the user can then begin to load utensils into the disinfecting chamber.

This sentence conveys different information and different steps in a process. Segmenting these items to understand what this sentence says can be challenging—particularly for nonnative English speakers. However, authors could convey the same ideas in multiple, short sentences such as

To start the disinfecting process, you must

- First, press the <START> button to begin the cleaning sequence. A green "on" light should then light up to indicate this process has begun.
- Next, the user should load the utensils into the disinfecting chamber.

Such structure splits each idea into its own sentence. It also formats information to present items in a sequence. Finally, it allows users to view each central action independently of others. Instructors need to emphasize that this structure makes it easier for nonnative English readers to identify the tasks to perform in relation to the process described here (Johnson, 2016; Kohl, 2008).

### Pedagogical Applications

In terms of teaching these concepts to students, instructors should focus on editing practices to address structure. In particular, instructors should focus on the following approaches:

### Passive Voice

For editing students, the key approaches for international audiences involve

- Identifying uses of passive voice within the text.
- Attempting to restructure passive sentences into active voice structure.
- Providing (if missing) prospective subjects/actors for these revised sentences.

### Noun Strings

To address the problem of noun strings, editing students should focus on

- Identifying passages where more than two of the same item (e.g., nouns) appear in a sequence.
- Reviewing the passage to identify the noun/subject and related descriptors.
- Revising the text to clarify the descriptive role of certain text.

### Complex Sentences

Addressing complex sentences could be the most challenging international structural aspect for students to grasp. To help with this process, instructors should focus on teaching students about

- Identifying places where compound, complex, and compound-complex sentences appear.
- Determining how many distinct ideas such sentences cover.
- Splitting each idea into its own simple sentence.
- Structuring the presentation of sentences to note the relationship of ideas.

Such editing practices address structural issues at the phrase and the sentence level. They also enhance the usability of content for international audiences. Once students gain familiarity with these practices, they could become some of the easier-to-implement practices in international editing situations.

## Expressions—The Methods We Use To Convey Meaning

Cultures often create expressions to convey a particular meaning in a way that seems inconsistent with the words used (Boers, 2008; Erfesoglou, 2017). In fact, when viewed literally, the words and phrases used might seem nonsensical (e.g., to "hit a home run during the business meeting"). Additionally, a culture might use particular terms to note certain attributes an individual, object or situation might have (e.g., "Jan was quite the fox to address that difficult business situation that way"; Horton, 1994). Such comparisons, however, are often specific to the cultures using them (Boers, 2008; Erfesoglou, 2017).

In other cases, cultures could use systems for representing numeric information for date, time, and magnitude, and these conventions can be specific to a given group (Frischknecht, 2015; Rodgers, 2017; St.Amant, 1999; Writing for translation, 2018). Such factors can cause confusion when communicating internationally. These elements thus become items technical editors must look for when reviewing materials for global audiences. Such factors often fall into certain categories: idiomatic expressions, metaphoric references, and numeric representations. In teaching this topic, editing instructors might focus on these areas by discussing them in terms of the ideas overviewed here.

### *Idiomatic Expressions*

An idiomatic expression is a phrase used to convey an idea that is different from what the related words literally say (Boers, 2008; Erfesoglou, 2017). The idiomatic expression "It's raining cats and dogs" does not literally mean these animals are falling from the sky like raindrops. Rather, a particular English-speaking culture associates a certain nonliteral meaning with this expression (i.e., it is raining forcefully). Yet, unless an individual learns the nonliteral meaning for that expression, that person has no way to interpret it as anything other than literally. Idiomatic expressions can therefore be among the most confusing and problematic items technical editors must identify and address.

This factor is the central concept instructors need to impart when teaching students editing for international audiences (LinguaSoft, 2015; Norvet, 2017; Yunker, 2003).

For nonnative English readers, a major challenge is identifying when writers are using an idiom, for they can be deeply ingrained into how cultures convey ideas. Idiomatic expressions such as "it is raining cats and dogs" are relatively easy to identify. Others, like to "break down a problem" in order to "unpack the bugs creating glitches in the system" might be less obvious. Technical editors therefore need to be diligent when reviewing content for idioms as the editor's familiarity with them could result in missing certain idiomatic expressions.

To address idiomatic items, the technical editor must first identify them. (This process might involve rereading a text several times in order to note less obvious, more commonly used idioms.) Next, the editor should determine if a particular idiom is needed, and, if not, delete the related expression. (It's one thing to revise "It's raining cats and dogs" and another to revise "It rains rather forcefully here on a daily basis; it seems like it's always raining cats and dogs.") If the related idea is essential to comprehension, the editor needs to revise the text to state the same concept by using terms that convey the literal meaning represented by the words in a phrase (e.g., using "It's raining forcefully" instead of "It's raining cats and dogs"). While these ideas seem simple, editing instructors need to emphasize how difficult this process can be in practice. They also need to emphasize the importance of doing multiple readings of a text to identify such factors.

## Metaphoric References

Metaphors generally involve comparisons based on the attributes the members of a culture associated with an item (Horton, 1994; Metaphoric Thinking, n.d.). To say "Pat is an owl when it comes to knowledge" draws on a cultural association that owls have certain traits (e.g., wisdom) seen in the person or item being compared with an owl (i.e., Pat is also wise). These associations are not universal (LinguaSoft, 2015; List, 2007; Rodgers, 2017). Rather, the use of such conventions can confuse individuals from other cultures. (Per Horton, 1994, owls, for example, are associated with stupidity in other cultures.)

Editing students need to understand that addressing metaphoric aspects is similar to addressing idioms in international environments. The technical editor needs to review the related content to identify areas where such comparisons are made. Next, the editor must determine if such comparisons are essential to conveying meaning in the related context. If so, the editor needs to revise wording to convey the related idea literally (e.g., "The individual is very wise" vs. "The individual has the wisdom of an owl") or to remove it entirely if not needed to convey information. Again, such concepts can seem quite simple. The editing instructor therefore needs to emphasize the importance of multiple readings of a text to identify such items when editing for international audiences.

## Numeric Representations

One might claim that numbers are culture free. The numeric representation for "1," after all, is generally considered universal. Yet, how cultures use numbers to convey information is not a standard process. Rather, it can differ from culture to culture (Frischknecht, 2015; St.Amant, 1999; Writing for translation, 2018). Technical editing students need to understand such factors when reviewing materials to make them more usable in global contexts. Editing students also need to realize that dates, times, and magnitudes are items of particular importance in relation to using numbers internationally.

Most cultures have numeric formats for all parts of a date—year, month, and day. Most cultures also identify the first month of the year as 1 (January) and the last as 12 (December). The order in which cultures organize these items in a date can differ (Frischknecht, 2015; Writing for translation, 2018). Some cultures order dates as month, day, and year and other cultures order date information as day, month, year—so the third of January 2018 would be 1-3-18 in some cultures and 3-1-18 in others.

The same is the case for time. Certain cultures use a 24-hour clock to represent time while others use a 12-hour clock with "a.m." for prenoon times and "p.m." to indicate postnoon times. So, certain cultures will meet at 1:00 p.m. (or 1:00) and others at 13.00—both of which are the same time.

Cross-cultural differences can also occur with magnitudes (St.Amant, 1999; Shearer, 2012). Certain cultures use commas (,) to indicate larger numbers and periods (.) smaller ones. Other cultures use these markings of magnitude in inverse order. This factor means the number "one-thousand, three-hundred thirty-three" would be "1,333" to some but "1.333" to others. Similarly, the number "one and one third" would be "1.333" for some cultures but "1,333" to others.

In teaching this topic, instructors must train editing students to first identify places where numbers are used to convey information in a text. Students next need to determine a format for making that information accessible to as wide an international audience as possible. In the case of dates, this could involve spelling out the month and using the full calendar year (so, "1-3-12" would become "1 March 2012"). In the case of times, this could involve using both 12- and 24-hour formats to present times (e.g., "The meeting took place at 3:00 p.m./15.00"). In the case of magnitude, it can involve indicating the full value of a number to avoid confusion (e.g., "The value is 1,300.04.") or using alternative formats to note smaller vs. larger magnitudes (e.g., using "1-1/3" instead of "1.333"). In such situations, students need to make sure such formats are used repeatedly and consistently throughout a text to avoid confusion about the information they convey.

## Pedagogical Applications

When teaching these topics, instructors should focus on students learning the following strategies and practices.

### Idiomatic Expressions

While idiomatic expressions are among the most problematic items, they can also be relatively easy to address if editing students learn the process of

- Identifying where such expressions are used.
- Determining if the idea needs to remain in the text.
- Revising the text to

  o Delete such items (if not essential to meaning).
  o Reword items to convey intended meaning literally (if essential to usability).

### Metaphoric References

When teaching editing students about metaphoric references, instructors should also teach them to edit texts by

- Identifying where such expressions are used.
- Determining if the idea needs to remain in the text.
- Revising the text to

  o Delete such items (if not essential to meaning).
  o Reword items to convey intended meaning literally (if essential to usability).

### Numeric Representations

Teaching students international editing practices for numbers involves students learning the approach of

- Identifying places where numeric representations occur in the text.
- Spelling out the name of the month and presenting all four numbers in a year (e.g., 3 January 2011).
- Using comparative formatting for times (e.g., "2:00 p.m./14.00").
- Presenting smaller-order magnitudes with nonpunctuated formats if possible (e.g. "1-1/4").
- Using full numeric format to present large-order magnitudes (e.g., "1,300.00").

Such practices involve editing students applying ideas and approaches they have likely encountered before to a new (i.e., international) context.

## Additional Resources—The Materials We Use to Design Texts

Organizations might ask technical editors to develop resources to guide the design of texts for international audiences (Esselink, 2000; Yunker, 2003). Such resources often use the general ideas of editing for global English as a foundation for content-creation practices. This section examines three resources to cover when teaching international editing practices.

### Glossary

As noted, restricting the use of certain terms can reduce confusion and increase usability. For such practices to be effective, content creators need to have a common understanding of how they should use a given word within a text. One mechanism for addressing such factors is a *glossary* that establishes a common definition for core concepts (Esselink, 2000; St.Amant, 2013; Yunker, 2003). Technical editors might be tasked with developing or using such resources. For this reason, editing students need to have some understanding of glossary creation relating to international contexts.

In teaching this topic, instructors should note that glossaries can also facilitate translation (St.Amant, 2013; Yunker, 2003). Translators might be aware that certain terms have multiple meanings in the source language. This factor can lead to delays as the translator queries the author to confirm the meaning of words in a text. Glossaries that stipulate how the term is defined and used can remove the need for asking such questions and reduce the time needed to translate overall texts (St.Amant, 2013).

Students need to understand that technical editors should create glossaries that authors/content creators and translators alike can use to guide their work. The editor should therefore review source texts to determine what potentially problematic terms occur in a document. To connect this process to prior topics, instructors should have students develop glossaries by focusing on

- Terms that have multiple meanings (e.g., set).
- Terms that should be used in a particular way within a text (e.g., physician vs. doctor).
- Terms that are specialized and essential to the ideas discussed, but unfamiliar to a wider audience (e.g., the medical process *cricothyrotomy*).

Such a focus can help students conceptualize glossaries within the greater context of editing for international audiences.

The next concept to impart to students is that technical editors need to develop a glossary entry for each of these kinds of terms. Such entries should provide the information needed for effective content creation or translation. In teaching this topic, instructors should explain that each glossary entry should contain

- The term/word.
- The part of speech (e.g., noun, verb, adjective, etc.) the term plays in an entry.
- A definition of the term.
- A sample sentence displaying the use of the term as related to this specific definition (St.Amant, 2013).

In terms of formatting, an entry might look like this:

**set** (noun): A collection of common or like objects

Example: *Place all of the test tubes in one box to make a set of similar containers to transport to the laboratory.*

Editing instructors should also note that technical editors might include glossaries in materials for international audiences depending on the purpose and kind of document (e.g., a glossary section within a technical manual) (Kohl, 2008; Yunker, 2003).

## *Stylesheet*

Content creators and editors regularly use stylesheets to guide the creation and revision of documents (Rude & Eaton, 2011). Addressing such factors makes content more usable—particularly for a given audience. (Editing instructors will likely have raised this idea in other parts of an editing class.) These stylesheets can cover everything from the sizes and styles of font to use to the expected uses of certain terms. Editing students need to understand that these same ideas can be used to create stylesheets associated with international usability.

When developing such stylesheets, the focus should parallel the prior ideas on editing for global English to provide authors with information to guide how they create texts (Esselink, 2000; Flint, vanSlyke, Starke-Meyerring, & Thompson, 1999; Lange & Bennett, 2000; Yunker, 2003). To connect this concept to international editing, instructors need to help students see the relationships between stylesheets and international content creation. Instructors can do this by focusing on certain points. They should, for example, note how to use vocabulary in specific ways (e.g., use of words). They should also explain the kinds of expressions to avoid (e.g., idioms and metaphoric meaning) when creating a text for international audiences. These guides can also provide instruction on how to represent certain ideas (e.g., for presenting dates and times) to avoid confusion in international contexts (Esselink, 2000; Flint et al., 1999; Yunker, 2003).

To help students conceptualize such practices, the instructor should explain that these approaches involve the international editing suggestions noted earlier. In this case, one is providing this information in the form of instructions others can use when creating or editing texts (e.g., "Avoid passive voice; instead, use active voice for all sentences" or "Revise all passive voice sentences to active voice sentences") (Esselink, 2000; Flint et al., 1999).

Students should also learn that an important item to include in such stylesheets is a glossary (Lange & Bennett, 2000; Yunker, 2003). For authoring/content-creation stylesheets, this glossary would identify terms authors should only use in certain ways within a document. (This approach helps avoid the wording-related pitfalls noted earlier.) For editors, this glossary would help to

- Confirm that content creators have used key terms according to glossary expectations (i.e., with a consistent meaning across an overall text).
- Determine if other terms used in a text might be problematic because they contradict glossary entries (e.g., using *doctor* in a text, but the glossary identifies *physician* as the term to use when conveying this idea).

The objective is to teach students that technical editors should view stylesheet creation as connected to glossary development and to do both in conjunction with each other.

## Controlled Language

Controlled language represents an acute application of using stylesheets to create content for international audiences—including translators. A controlled language is one that conveys ideas via a restricted vocabulary—essentially, a vocabulary in which terms are exclusively used to convey one idea (Esselink, 2000; Lockwood, 2000; Yunker, 2003). Controlled language also often involves using sentences of limited length, format, and content (e.g., using only active voice, simple sentences comprised of 10 or fewer words; Esselink, 2000; Lockwood, 2000; Yunker, 2003). Ideally, these restrictions limit the potential misinterpretation of terms and mitigate the misunderstanding that could arise from more complex writing styles. Controlled language can thus serve as a mechanism for creating or editing texts for international users. For this reason, instructors should include the topic of controlled language when teaching students about editing in international contexts.

A core idea for instructors to emphasize is that the limitations of controlled language mean that it generally cannot be used across a wide range of topics and contexts (Esselink, 2000; Lockwood, 2000; Yunker, 2003). Rather, it generally reflects the needs of a specific industry—for example, the aviation industry where individuals use Aviation English to communicate internationally (Wojcik & Holmback, 1996). The controlled language might also reflect a particular company and the products it creates for international consumers. Such is the case for Caterpillar Fundamental English, a controlled language that the heavy equipment producer Caterpillar uses to convey information about its specific products to global audiences (Lockwood, 2000). Editing students need to understand that it is the specific focus on a particular industry or organization that makes a controlled language relatively easy to implement in a limited/restricted setting. A controlled language could also involve a glossary that defines certain industry- or company-specific terms that one must use when communicating information about particular topics (Lockwood, 2000).

When teaching this topic, instructors should note that controlled languages can affect international editing practices in relation to audience and use. First, a controlled language can serve as the focus of the style/editing guidelines technical editors must use when reviewing and revising documents. In such cases, the technical editor might need to review content created in a controlled language to determine how effectively it conforms to the related requirements (e.g., usability). In other cases, the technical editor might need to create such guidelines based on the needs of a given client (i.e., audience). Ideally, the client would have an existing set of content creation guidelines that technical editors could develop into the editing guidelines used to review controlled-language texts. (Developing such guidelines from scratch can be a complex and time-consuming process involving management, content creators, editors, and audiences/content users.)

In such situations, glossaries are an important resource for technical editors. Here, glossaries can guide how one reviews materials in terms of the words they contain. Glossaries can also guide how the content creator uses terms in a given text. When developing controlled language guidelines, the editor might identify the terms that will become core verbal elements in a controlled language vocabulary. The editor would then create an associated entry—including the expected meaning for the term and related requirements for when to use it. Drafting such glossary entries can be a complex process involving many parties (Esselink, 2000; Lockwood, 2000; Yunker, 2003). Technical editors should thus consider the dynamics associated with crafting a controlled language glossary/vocabulary before agreeing to such projects. By examining these topics this way, instructors can help editing students to

- See how such international editing practices are interconnected.
- Understand how these practices are founded on core technical editing concepts.

These factors can help students perceive international editing as an extension of the technical editing work they do and of the concepts that guide overall editing practices.

## Conclusion—The Central Points to Remember

The interconnected nature of the global economy means organizations will need to increasingly think of their communication activities in terms of international contexts. Effectively addressing such situations involves understanding how certain linguistic and stylistic elements can affect user comprehension and the usability of different materials. Technical editing students need to understand and train for such contexts to be successful in today's global workplace. Editing instructors can address this need by teaching students how to identify and revise certain elements in order to develop materials for international audiences. Doing so involves teaching students how editing for global English, glossary building, and stylesheet creation help technical editors enhance the usability and translatability of content for different users.

This chapter presented foundational concepts for introducing editing students to this area of technical editing. By building upon the skills and tasks noted here, editing instructors can better prepare students to contribute to the content creation, editing, and international communication practices of a range of organizations. As the need for such materials—and the skills required to produce them—grows, so do the ways in which technical editors can contribute value to an organization's efforts to "go global." Integrating such concepts into technical editing classes can thus enhance the opportunities for student success after graduation.

## Pedagogical Practicalities

To help students apply the information and practices noted in this chapter, instructors could integrate the following activities into an editing class. Such activities could be done during a class session or as a homework assignment in which students discuss their work during a subsequent class meeting. The central aspect of each of these suggested applications is that students discuss what they did for a given activity to gain a more complete understanding of how to implement the ideas noted in this chapter as a part of technical editing practices.

### Activity 1: Editing for Global English

The instructor could assign each student to review and edit a particular page of a website created by a U.S. entity for U.S. audiences (e.g., the website for the related institution at which the student is taking the class) in terms of the global English items discussed in this chapter. In making such edits, students would be required to

- Note the item that needs revision to address global English concepts.
- Propose a suggested revision to address global English factors.
- Provide a comment in which the student explains the reason for the suggested edit.

This approach helps students practice the use of global English concepts to guide international editing practices. The related application of ideas also helps them better understand (via articulating the justification for a given edit) the reason for making such edits. Instructors could have students repeat this activity with different kinds of web pages (e.g., local or state government agencies or local businesses) to have students practice editing for global English with different kinds of entities. Doing so could help students understand if and how the kind of client/content provider (e.g., educational institution, government agency, private business) might affect the items one needs to address when editing for global English.

## Activity 2: Glossary Creation

The instructors would have students review a website—or websites—for a local or regional business or organization and create a glossary that editors or translators could use to revise the related text for international audiences (either global English readers or translators). In creating such glossaries, students would need to

- Identify terms they believe require glossary entries for international audiences or for translators.
- Provide an explanation of why each term merits a glossary entry.
- Craft a four-part glossary entry (per the criteria noted in this chapter) for each of these terms.

This mix of identification, explanation, and entry creation can help students apply ideas and engage with them (via explanations for term selection) in a way that helps them better conceptualize and apply the concepts they learned in relation to this topic.

## Activity 3: Applying Stylesheets

During one class meeting, the instructor could work with the students to develop a stylesheet that the related program or department (that offers the class) can use to create information according to global English factors. Once created, each student could use that stylesheet to review a page of the program's, department's, or unit's (e.g., college's or school's) website and revise it according to the parameters of that stylesheet. In making such edits, students would need to

- Note the item that needs revision to address global English concepts.
- Propose a suggested revision to address global English factors.
- Explain how that proposed revision meets a particular stylesheet item.

After the students have completed this activity, they would share the results of their work—suggested revisions and reasons for revisions—with the other members of the class. The class would then use this information to discuss if the stylesheet created for the program/department requires revision, or if it is effective as it is. (If revision is needed, the nature and need for such revision could be discussed and decided upon during that meeting.) Through this approach, students learn the nuances and practices of creating and applying stylesheets to address international audience needs. Students also learn how review and revision is an important component of developing such stylesheets.

## References

Bartels, A. (2017, Oct. 18). Global Tech Market Will Grow by 4% in 2018, Reaching $3 Trillion. *Forbes*. Retrieved from https://www.forbes.com/sites/forrester/2017/10/18/global-tech-market-will-grow-by-4-in-2018-reaching-3-trillion/#101ff69412c9

Boers, F. (2008). Understanding idioms. *MED Magazine*, February. Retrieved from http://www.macmillandictionaries.com/MED-Magazine/February2008/49-LA-Idioms.htm#4

Corbin, M., Moell, P., & Boyd, M. (2002). Technical editing as quality assurance: Adding value to content. *Technical Communication, 49*(3), 286–300.

Emerging markets: Statistical profile. (2018). *NationMaster*. Retrieved from http://www.nationmaster.com/country-info/groups/Emerging-markets

Erfesoglou, L. (2017). Idioms: A linguistic journey across cultures. *Cultural Awareness International*. Retrieved from https://culturalawareness.com/idioms-linguistic-journey-across-cultures

Esselink, B. (2000). *A practical guide to localization*. Philadelphia, PA: John Benjamins.

Flint, P., vanSlyke, M. L., Starke-Meyerring, D., & Thompson, A. (1999). Going online: Helping technical communicators help translators. *Technical Communication, 46*(2), 238–248.

Frischknecht, S. (2015). Writing for translation: 10 translation tips to boost content quality. *Lionbridge*. Retrieved from http://content.lionbridge.com/writing-for-translation-10-expert-tips-to-boost-content-quality

Guidelines for writing global English. (2018). *EdX*. Retrieved from http://draft-edx-style-guide.readthedocs.io/en/latest/global_English.html

Horton, W. (1994). *The icon book: Visual symbols for computer systems and documentation*. New York, NY: John Wiley & Sons.

Johnson, M. R. (2016). How global is your English: 8 ways to keep it simple and save big. *The Content Wrangler*. Retrieved from https://thecontentwrangler.com/2016/02/02/how-global-is-your-english-8-ways-to-keep-it-simple-and-save-big

Kaney, T. (2018). 13 tips to remember when writing for translation. *The GEO Group*. Retrieved from https://www.thegeogroup.com/13-tips-to-remember-when-writing-for-translation

Kohl, J. R. (2008). *The global English style guide: Writing clear, translatable documentation for a global market*. Cary, NC: SAS.

Lange, C. A., & Bennett, W. S. (2000). Combining machine translation with translation memory at Baan. In R. C. Sprung (Ed.), *Translating into success: Cutting-edge strategies for going multilingual in a global age* (pp. 203–218). Philadelphia, PA: John Benjamins.

LinguaSoft. (2015). *Tips on writing for translation.* Retrieved from https://www.mtm linguasoft.com/wp-content/uploads/MTM-LinguaSoft-tips-for-writing-for-translation.pdf

List, D. (2007). How to write global English: Some principles for clear writing. *Audience Dialogue.* Retrieved from www.audiencedialogue.net/english2.html

Lockwood, R. (2000). Machine translation and controlled authoring at Caterpillar. In R. C. Sprung (Ed.), *Translating into success: Cutting-edge strategies for going multilingual in a global age* (pp. 187–202). Philadelphia, PA: John Benjamins.

Lyons, D. (2017). How many people speak English, and where is it spoken? *Babbel Magazine.* Retrieved from https://www.babbel.com/en/magazine/how-many-people-speak-english-and-where-is-it-spoken

Maylath, B. (2011). Editing for global contexts. In C. D. Rude and A. Eaton (Authors), *Technical editing* (5th ed.; pp. 300–317). Boston, MA: Longman.

Medical devices regulations, SOR/98-282. (2017). Canadian Legal Information Institute. Retrieved from www.canlii.org/en/ca/laws/regu/sor-98-282/latest/sor-98-282.html

Metaphoric Thinking. (n.d.) *Mind Tools.* Retrieved from https://www.mindtools.com/pages/article/newCT_93.htm

Morningside Translations. (2016). Multiple market product launches: How multi-national companies use SIMSHIPS. *ITProPortal.* Retrieved from https://www.itproportal.com/features/multiple-market-product-launches-how-multinational-companies-use-simships

Norvet, A. (2017). What is global English? *United Language Group.* Retrieved from http://daily.unitedlanguagegroup.com/stories/editorials/global-english

Official languages of the EU. (2018). *European Commission.* Retrieved from http://ec.europa.eu/education/official-languages-eu_en

Passive voice and active voice. (2001). *The Mayfield handbook of technical and scientific writing.* Retrieved from http://www.mhhe.com/mayfieldpub/tsw/pass-act.htm

Rodgers, J. (2017). Guidelines for writing for translation. *Web-Translations.* Retrieved from http://www.web-translations.com/blog/content-guidelines

Rogers, J. (2018). What's so bad about the passive voice? *The Rabbit Room.* Retrieved from http://rabbitroom.com/2018/03/whats-so-bad-about-the-passive-voice

Rogers, K. (2017). As the Earth feels ever smaller, demand for translators and interpreters skyrockets. *CNBC.com.* Retrieved from https://www.cnbc.com/2017/07/07/as-the-earth-feels-ever-smaller-demand-for-translators-and-interpreters-skyrockets.html

Romero, S. (2017). Spanish thrives in the U.S. despite an English-only drive. *New York Times*, August, 23. Retrieved from https://www.nytimes.com/2017/08/23/us/spanish-language-united-states.html

Rosenberg, M. (2017). What countries have English as an official language? *ThoughtCo.* Retrieved from https://www.thoughtco.com/english-speaking-countries-1435414

Rude, C. D. & A. Eaton. (2011). *Technical editing* (5th ed.). New York, NY: Allyn & Bacon.

St.Amant, K. (1999). International integers and intercultural expectations. *Intercom*, 5–27.

St.Amant, K. (2013). Understanding the process of language translation: A primer for medical writers. *American Medical Writers Association Journal, 28*(1), 3–8.

St. Germaine-McDaniel, N. (2010). Technical communication in the health fields: Executive Order 13166 and its impact on translation and localization. *Technical Communication, 57*(3), 251–265.

Shearer, K. (2012). Telling time around the world. *CultureWizard.* Retrieved from https://www.rw-3.com/blog/2012/02/different-ways-of-talking-about-time

Sprung, R. C., & Vourvoulias-Bush, A. (2000). Adapting *Time Magazine* for Latin America. In R. C. Sprung (Ed.), *Translating into success: Cutting-edge strategies for going multilingual in a global age* (pp. 13–27). Philadelphia, PA: John Benjamins.

Stein, L. (2015). What is simplified English and how can it reduce the cost of my translations? *Net-Translators*. Retrieved from https://www.net-translators.com/blog/reduce-translation-costs-with-simplified-english

The Acrolinx Team. (2015). Do you write in global English? *Acrolinx*. Retrieved from https://www.acrolinx.com/content-marketing/write-global-english

The passive voice: Good or bad? (n.d.). *WhiteSmoke*. Retrieved from http://www.whitesmoke.com/passive-voice-in-english

Ulijn, J. M., & Strother, J. B. (1995). *Communicating in business and technology: From psycholinguistic theory to international practice*. Frankfurt: Peter Lang.

Winninger, P. (2013). Foundations: The role of the technical editor. *TECH Whirl*. Retrieved from https://techwhirl.com/foundations-the-role-of-the-technical-editor

Wojcik, R., & Holmback, H. (1996). Getting a controlled language off the ground at Boeing. Paper presented at the 1st International Workshop on Controlled Language Applications, Centre for Computational Linguistics, Leuven, Belgium, March 26–27, 1996.

Writing for translation. (2018). *MailChimp*. Retrieved from https://styleguide.mailchimp.com/writing-for-translation

Yunker, J. (2003). *Beyond borders: Web globalization strategies*. Indianapolis, IN: New Riders.

# 9

# A FIELD-WIDE VIEW OF UNDERGRADUATE AND GRADUATE EDITING COURSES IN TECHNICAL AND PROFESSIONAL COMMUNICATION PROGRAMS

*Lisa Melonçon*

## Chapter Takeaways

This chapter

- Provides technical and professional communication program administrations (TPC PAs) and faculty the ability to ensure that their editing courses align (in ways that make sense locally) with national trends and approaches.
- Provides insights into how editing courses can be improved.
- Encourages a more in-depth engagement with how an editing course is defined and what should be included.

The editing course has been identified as a "core course" in both undergraduate and graduate degree programs (Melonçon, 2009; Melonçon & Henschel, 2013). While this course may be offered in a large number of programs currently, TPC PAs and faculty have little understanding of what is actually being taught in it. In similar ways to Chong's (2016) study on usability courses across programs or the more analogous study on capstones (Melonçon & Schreiber, 2018), this chapter seeks to provide an overview of undergraduate and graduate editing courses at the field-wide level. Programmatically, the broader view from across the field enables technical and professional communication (TPC) to look beyond individual program-specific case studies, which can set up the potential to design and/or to improve editing courses.

Thus, the goal of this chapter is straightforward: to offer TPC PAs and faculty insights on the current pedagogical approaches to the editing course. After an overview of the methods for data collection, I provide information from course catalogs, syllabi, course materials, and interviews with TPC PAs and/or faculty. Informative sections include course titles and descriptions, general approaches to the course, common assignments, crossover courses, and faculty staffing. The final section looks at implications for TPC including what is currently working in editing courses and areas for improvement.

## Methods for Collecting Data

I confined my analysis to institutions where the editing course is required as a part of a degree program. *Degree program* here means that a student can pursue a credential that will appear on her transcript, and degree programs include undergraduate degrees (both degrees in TPC and those degrees in something with an emphasis, concentration, or track in TPC), graduate and undergraduate certificates, minors, and master's degrees. I limited my analysis to institutions with required editing courses; required courses suggest a level of agreement among faculty about the importance of the course as necessary for the degree program. The limited scope also made data gathering and subsequent analysis more manageable.

Using a current list of degree programs from TechComm Programmatic Central (Melonçon, 2018), I located all the institutions that *require* an editing course no matter what type of degree program it is. (See Facts and Figures below for a complete breakdown.) Following the methodology of previous programmatic work, I started my investigation with information found in course catalogs (Melonçon & Henschel, 2013, pp. 46–47). The course titles, course descriptions, prerequisites, and credit hours were gathered, but it was clear additional information would be needed to fully understand the editing course and how it was being taught.

As a required course (discussed in more detail below), it is most likely that the course would be offered at least every other year, so I went to each institution's scheduling system and located who taught the course within the last two years. This type of attention to detail in the research study design ensures the most accurate information and also provides the opportunity to ask follow-up questions to the person who is teaching the course. While time-consuming, this approach to research study design is essential in gathering accurate and comprehensive data about courses. More so, gathering information in this way can lead to research results that can potentially be generalized, which is not something that can occur from generic survey requests to disciplinary listservs. (See Melonçon, 2018, for more information on research study design for programmatic research like this.)

To obtain a richer data set that included localized contextual data, in late fall 2017 and early spring 2018 I emailed faculty members asking if they would share their course materials. I emailed all the graduate level ($n = 19$) and crossover course ($n = 11$) faculty. Because of the large number of undergraduate degree programs ($n = 105$), I selected half of the institutions to contact based solely on the ease of finding online information about who was teaching the course. That is, some institutions' websites are so painfully difficult to navigate that some information was never located. Thus, when I located contact information for half of the undergraduate degree institutions, I went with that list because previous research indicated that this number of contacts would generate adequate data for analysis and a reliable set of data from which broader generalizations could be made. I sent out requests to this list and asked for syllabi, assignments, reading lists, and/or anything that would help the field understand the pedagogical approach to the course. I intentionally did not define pedagogical approach nor did I put stipulations on the request for information.

I received information from 40 schools: 11 graduate (69% return rate); 4 crossover courses (36% return rate); 26 undergraduate programs (60% return rate). In some cases, I completed an asynchronous interview with faculty (St.Amant & Melonçon, 2016, p. 349) that garnered additional institutional context for the course materials that were provided. All of the specific identifying information has been removed based on Institutional Review Board (IRB) stipulations[1] and/or my own approach to research practices, which aims to put the focus on the curricula and information rather than on individual programs. Therefore, data reported are referred to by Carnegie classification (see http://carnegieclassifications.iu.edu for a full description of classification types). For example, I may discuss the editing course from an R2 institution (higher research activity).

## Facts and Figures

In this section I provide basic facts and figures about editing courses and the data that was gathered.

### Number of Required Courses

What follows is the number of required editing courses and the type of degree program in which they are required. The difference in the number of degree programs and the number of faculty contacted can be attributed to several institutions that require an editing course for multiple degree programs.

- 105 undergraduate degree programs (at 86 institutions) that require an editing course:
    - 29 TPC degree programs (out of 74 degree programs in TechComm Programmatic Central);
    - 31 emphasis degree programs (out of 125 degree programs in TechComm Programmatic Central);
    - 26 minors (out of 141 degree programs in TechComm Programmatic Central);
    - 19 undergraduate certificates (out of 67 degree programs in TechComm Programmatic Central).

- 19 graduate degree programs (at 16 institutions) that require an editing course:

  ○ 13 MA/MS degree programs (out of 111 degree programs in TechComm Programmatic Central);
  ○ 6 certificate programs (out of 52 degree programs in TechComm Programmatic Central).

  Crossover courses break down as follows:

- 20 degree programs (at 11 institutions) that require an editing course:

  ○ 7 MA/MS degree programs;
  ○ 3 graduate certificates;
  ○ 3 undergraduate TPC degree programs;
  ○ 4 undergraduate emphasis degree programs;
  ○ 1 undergraduate certificate;
  ○ 2 minor degree programs.

Table 9.1 provides the number of undergraduate courses for each academic level. Crossover editing courses are at the senior-level undergraduate level and the introductory graduate level.

## Course Titles

As the outward-facing information that students and other stakeholders see, course titles (and descriptions) are important institutional and programmatic documentation. Unlike other courses, however, the editing course is a bit more straightforward in being able to succinctly describe what the course contains based on its title. The data gathered indicates field-wide commonalities in course titles across the United States. The information here includes all of the required editing courses in all types of programs.

- All graduate courses and crossover courses include *editing* in the title.
- 10 of 19 graduate courses include *technical*.
- 15 out of 19 crossover courses include *technical*.

TABLE 9.1 Academic level of undergraduate editing courses.

| Academic level | # courses |
| --- | --- |
| 100: freshman | 1 |
| 200: sophomore | 10 |
| 300: junior | 42 |
| 400: senior | 31 |
| 500: advanced | 2 |

Undergraduate courses fell into the following types of course based on the titles:

- Editing—20.
- Publishing or publication (e.g., editing and publication)—9.
- Professional (e.g., professional editing)—10.
- Style (editing for style and clarity)—19.
- Technical (e.g. technical editing)—28.

Additionally, only four undergraduate editing courses had *grammar* in the titles, while all but two undergraduate courses had *editing* or *edit* in the course title.

Based on the titles of the editing courses, the field seems to agree that *editing* is a key marker for students, faculty, administrators, and those in industry who would review transcripts or curriculum. However, at the undergraduate level the inclusion of *style* or *publishing* indicates that the courses do have different foci.

## Course Descriptions

Understanding the limitations of course descriptions as a form of institutional documentation that is often changed by the instructor, course descriptions still provide important insights into the editing course at the field level for that very same reason. That is, they are the institutional description of the course that is outward-facing to students.

What follows are representative samples from course descriptions across both undergraduate and graduate editing courses. They are listed here together since there was no discernible difference in the course descriptions at the different levels.

- Grammar & mechanics, copyediting, style, organization, graphics, electronic editing, professional concerns. Major portfolio assignment. Essential for students planning to work as technical editors. (Research 1, undergraduate course.)
- Students will examine the general principles and practices of sound editing of and for writing in the workplace. The course prepares students in all aspects of editing documents, from proofreading for surface errors to ensuring appropriate content, organization, visual elements or components, and usability, needed for editing tasks in the workplace. (Master's Medium, undergraduate course.)
- Theoretical and practical grounding in professional workplace editing as an evolving profession. Design, edit, and manage complex workplace documents using both manual means and industry-standard software. (Master's Large, undergraduate course.)

- Study of advanced technical communication situations such as formal reports, grant proposals, and professional articles, and extensive discipline-specific professional level practice in these forms. Study of general editorial techniques in formats, graphics, and layout and design methods in technical publications. (Master's Large, undergraduate course.)
- Proofreading, copy-editing, comprehensive editing. Students primarily use electronic editing methods. Editor's responsibilities, relationship to writers, roles within an organization, style guides, ethical choices. Editing in global setting. Editing/style for visual design and online documents. (Research 2, graduate course.)
- Examines complex roles editors assume in creating technical and nontechnical documents. Principal components include working with substance of documents, mediating the writer–reader relationship, and exemplifying the application of rhetorical theory in editing. (Research 2, graduate course.)
- Editing the content, organization, format, style, and mechanics of documents; managing the production cycle of documents; and discovering and learning computer and software applications for technical editing tasks. (Master's Large, crossover course.)

The common characteristic, no matter the emphasis in the description, is practice-based experience. In addition, course descriptions emphasize that the course covers the full range of editing practices from line edits and grammar to larger issues of aligning purpose and audience.

In a recent conversation in the LinkedIn Technical Communication Forum, a question was raised about what academic programs actually include in a course because the original poster was frustrated by the course descriptions she could find online. As I found out, editing courses are much more involved and include more diverse work than what is found in the course descriptions. Admittedly, I understand the difficulties of getting catalog information changed or updated and the delays in that information appearing online. However, as Melonçon and Schreiber (2018) pointed out,

> the external views of the field will remain limited unless TPC PAs and faculty take on this rhetorical challenge and find better ways to write about and document the processes and practices that are actually occurring in our courses and programs and foreground connections between academic concepts and TPC knowledge work.
>
> (p. 329)

This is definitely true in the case of the editing course. As the descriptions above indicate, much of the language being used to describe the course needs to be edited to be more concrete with a clear view of what students will be doing and to what end. The next section, however, does shed additional light on the actual practices and approaches that occur in the editing course.

## Approaches to the Course

The information in this section about approaches to the editing course is based on course materials received from faculty teaching the course. In total, I received information from 40 institutions: syllabi with schedules of topics and assignment descriptions/assignment sheets, as well as additional information that placed the course in a local context or provided additional context for the course itself.

### Topics

I use *topic* here to mean the weekly goals set forth in a syllabus. The topic description indicates to students what the focus of the reading and classroom exercises will be for a particular week, and when the topics are read together across the term they suggest what instructors—and programs—feel is valuable for students to learn in a course. What follows are the most common topics found and a brief description of what they cover:

- Fundamentals of editing: Spelling, grammar, punctuation, style.
- Types of edits: Copyediting marks both on paper and electronic.
- Practical editing: How to work with writers, types of editing, history of editing, working with clients; the process (publishing).
- Tools: Style guides and computer tools (e.g., Microsoft Word's track changes and Adobe Acrobat's editing features).
- Comprehensive editing: Larger order concerns around purpose and audience are being attended to as well as issues of organization and content.

One can see how these primary topics could be spread over an 8-, 10-, or 15–16-week term to provide students with both the skills and approaches to editing.

### Assignments

The most common types of assignments could be placed into the following categories:

- Quizzes and exams
- Style exercises
- Comprehensive and/or client-based projects.

### Quizzes and Exams

Across the board all institutions required some type of quiz or weekly exercise/drills to ensure that students were gaining proficiency in certain editing tasks. As Boettger (2014) has found, editing tests are a common part of hiring practices for technical and professional writers, which makes this good practice for students. These assignments were mostly confined to fundamental editing around grammar, spelling, punctuation, and ensuring the consistency of these items throughout a document. These also included quizzes or tests around copyediting marks. This category of assignment also included midterm and/or final exams, which are another way to gauge student proficiency around certain topics.

## Style Exercises

Style exercises typically fell into one of three approaches:

1. Analysis of style guides so students could see the rhetorical nature of them, as well as their functional use to make decisions for consistency.
2. Creation of a style guide.
3. Application of a style guide to a small-scale editing exercise.

The emphasis on style guides demonstrates the need for students to understand that writing and designing is usually driven by a set of standards that is particular to an organization. The style-guide assignments most often indicated the disciplinary or industry-specific preferences for certain styles as well as some of the differences between print and online styles. Learning that these standards can change across organizations and across style guides themselves is an essential skill for students to combine with understanding the nuance and complexity of audience and purpose.

## Comprehensive and/or Client-Based Editing Projects

The comprehensive editing project is designed to have students start the editing process thinking about audience, purpose, and use; and then move on to issues of design, consistency, organization, clarity, and support; and then to sentence-level issues of grammar, etc. The comprehensive editing projects were the most diverse in subject matter or requirements, but they were consistent with the aims and outcomes of the project. Some examples of comprehensive editing projects included websites, manuals, procedures, reports, academic articles, and newsletters.

While much of the course up to this point was instructor-driven, the comprehensive editing project was typically student-chosen and also had the option of working in collaboration with another student(s). Many of the comprehensive or client projects also had the requirement that the student editor(s) needed to correspond or interact with the author(s) of the document so they could gain practice in working with authors and learning the process of editing. Several of the assignment sheets had similar language when they stressed that the projects needed to be "real," "live," or "from an actual author or organization."

The emphasis on client-based projects is a hallmark of TPC education and has been found to be a characteristic of other courses within the TPC curricula (Melonçon & Schreiber, 2018).

## Metacognitive Work

A large number of institutions required students to provide an additional document that explained or provided a rationale for their decisions. These supplemental parts to assignments are worth noting since they do such important pedagogical and learning work. The "reflective" or "explanatory" memo requires students to discuss their decisions in light of course materials and in light of the goal for the assignment. This type of metacognitive work ensures that even in the applied, client-focused projects students are being asked to think critically about the work and knowledge production of editing. More so, incorporating metacognitive aspects for assignments situates students to better transfer their editing knowledge to other situations, which is vital since some form of editing is crucial to the role of the technical and professional communicator in any organization.

## Implications for TPC Programs

In this final section I overview what editing courses are doing right, where they can be improved, how labor issues impact curricula, and, finally, I address the biggest question that was raised with this research, "what is editing?"

## What We're Doing Right

What was striking about the course materials was the attention to having students focus on audience, purpose, and language. These three facets remain key to any editing project no matter what form it may take, so even the digital and video content that Lang and Palmer (2017) advocate for could be improved by an editor who has a keen focus on audience, purpose, and language. It's important to remember that no program can ever be agile enough to keep up with every new trend in industry. It's impossible to do that for all sorts of institutional reasons from policies and procedures for updating curricula to hiring faculty with needed skills. But what has made TPC programs vital and sustainable for the last 60 years is their ability to ground students in transferable skills that move across modes and changes in technology. The most important of those skills is understanding that clear, concise, and (mostly) error-free writing is connected to a specific purpose and audience and must be designed and delivered in particular ways. Much of the curricula's pedagogical approaches are connected to underscoring these primary tenets of technical and professional communication.

## Editing Fundamentals

Moreover, the editing course is providing students with editing fundamentals—the process of editing, grammar, spelling, and other lower-order and higher-order concerns (explained above). This finding aligns with Lang and Palmer's (2017) suggestions for improving the editing course and with the findings from Kreth and Bowen's (2017) survey of working technical editors. The fundamentals of editing are also being discussed within programs from a variety of perspectives that include that of an independent editor (much like the ones who participated in the Kreth and Bowen survey), the author as editor, and editing as part of a collaborative writing process that crosses over work divisions. Fundamental skills in editing are still a key part of the work of a technical and professional communicator's job and need to remain in curricula moving forward.

There seems to be a disconnect between the skills in the limited number of job ads posted by Lang and Palmer (2017), which were from a decidedly limited view based on the Society for Technical Communication (STC), and the richer and more diverse picture of what a technical editor is across fields from information found in the work of Kreth and Bowen (2017). By expanding the view of a practicing technical editor beyond the STC, Kreth and Bowen's (2017) list of what editors do is in conflict with the perceived needs as outlined by Lang and Palmer (2017). Based on the information from the 250+ respondents from 12 different professional organizations of editors, Kreth and Bowen's (2017) data shows that rarely, if ever, do editors work in audio or video. The types and kinds of editing (see Tables 2 and 3, pp. 242–243) they do is much more in line with the types and kinds of assignments being given in TPC programs.

## Connecting to Workplace Practice

As briefly noted in the Assignments section above, the majority of course materials referenced a direct connection between the course and the practice of editing in the workplace. One way they help to make this connection is through assignments that encourage working on projects for real authors or organizations (rather than classroom-generated scenarios that can still provide important and relevant practice). Technical and professional communication has long advocated for client-based projects throughout the curriculum (e.g., Dubinsky, 2002; Youngblood & Mackiewicz, 2013) because "[c]lient-based writing classes may function as effective sites for bringing together the standards of school and work, as well as writing and technical fields" (Taylor, 2006, p.112). All institutions should require students to do this sort of practice-based work. The positive aspects of the editing course ensure that it not only meets the goals of the TPC programs, but that it also appeals to "students who are uninterested in majoring in technical communication but who might be attracted to complementary minors or certificate programs" (Kreth & Bowen, 2017, p. 254).

## Room for Improvement

While TPC programs are doing much right in their editing courses, there is room for improvement or room for asking questions about current practices. In this section, I focus on improving the editing course by thinking about whether the current course is too narrowly focused, by questioning pedagogical approaches, and by considering what the differences are between undergraduate- and graduate-level courses, and the inclusion of important topics that are not currently prevalent in editing courses.

## Too Narrow a Focus

Lang and Palmer (2017) assert that technical editing courses in higher education need to be revamped because they are too situated in "textually based processes" (p. 298). Their study, however, was limited in scope and focus since they only viewed three courses in higher education and a smattering of job ads. Therefore, their findings are overstated based on the evidence they provided, and one could say that the study was reverse-engineered to prove the point they wanted to make. Even with this substantial problem, I cannot argue with their biggest concern that the current iterations of editing courses are narrowly focused. However, Lang and Palmer (2017) imply that the current version of the editing course is a complete waste of time, and, based on the evidence in the actual courses and the information from Kreth and Bowen's (2017) well-designed study, I would not draw the same conclusion. More than anything, faculty need to understand that our ongoing reliance on poorly designed research that relies on limited evidence (Melonçon, 2018; Melonçon and St.Amant, 2018) places the field at a disadvantage in actually building and sustaining TPC programs.

A large portion of the courses spent a good deal of time on the fundamentals. While this is a positive aspect, it also deserves additional scrutiny as being something that is enabling editing courses to remain too narrowly focused. For example, a crossover course at a baccalaureate institution spends eight weeks on copyediting, terminology, practicing copyediting for correctness in hardcopy and electronic, and then editing for syntax and structure, which culminates in the midterm exam in which students put these fundamental editing skills into practice. Only three weeks are focused on a comprehensive editing project. The other three institutions where they use a crossover course and whose materials I received did not look much different from this model because they respectively spent five, five, and 11 weeks of a full semester term on fundamentals. Crossover courses are not unusual, with graduate courses spending roughly 50% of the academic term on fundamentals of editing.

While I do understand the need to provide students with the fundamentals of editing and appreciate the different ways, from quizzes to short practice assignments, that move students through these skills, I do wonder if the emphasis on fundamental editing is limiting or too narrow for the current workplace and whether we should be spending more time within a term on comprehensive editing and client-based projects. The latter of which also provides students with practice in working with clients/others and more generally in working in virtual environments.

## Pedagogical Approach

The course materials I received were a mix of face-to-face and online courses with almost every course relying heavily on an institutional learning management system (e.g., Canvas). This meant that, almost universally, at both the graduate and undergraduate level, a large percentage of the student grade was contingent on discussion posts or reading posts. I do understand that this is a common pedagogical practice as a means to ensure that students are reading the material, but it struck me as odd that, in what should be a practice-driven course, so much of the pedagogical focus would be on posts to a discussion board.

Since I recognize that the materials I received are only a slice of the pedagogical approach, I would like to raise the question: Is this the best pedagogical approach? And a related concern: Is this the most effective way to ensure that students are doing the reading and such that they need to apply to practice, specifically for such an applied course as editing?

Another aspect of the pedagogical approach that I found disconcerting (and I acknowledge this could just be me) was the heavy reliance on quizzes and exams. Admittedly, I can see how a variety of quizzes could be tied to editing drills and short exercises to gauge how well students are learning certain aspects of editing, but, as research in writing has shown, grammar is best learned in context (see Crovitz & Devereaux, 2017; e.g., Williams, 1981). For example, in over 90% of the courses that I examined, their materials had a grammar quiz. Having received a number of the actual assignment sheets for the grammar quiz, these were actual quizzes mostly devoid of contextual frameworks that might include a short scenario to help understand purpose and audience. I do understand that the purpose of the grammar quiz is to ensure that students themselves understand grammar in order to be able to find these errors in documents. However, I do have to wonder if a more effective pedagogical approach would be to place these sorts of checks within a broader context, such as the actual editing of shorter documents that had the same sort of errors found in the quizzes.

Technical and professional communication has always been a leader in pedagogical innovation with its emphasis on theory to practice, technological incorporation into pedagogy, and incorporation of experiential learning opportunities to name but a few ways. Thus, it struck me that the editing course materials and approaches as evidenced in the syllabi and other materials seem to be taking part of the context out of editing. So I will end this section on pedagogical approaches with the simple plea for TPC PAs and faculty to seriously consider new and innovative ways to "test" whether students are getting important concepts that do not eliminate the rhetorical context and reduce editing to the popular-culture notion of the editor as someone who is only concerned with grammar.

## Differences Between Undergraduate and Graduate

Lang and Palmer (2017) situated their view of the editing course at the graduate level, and I have provided more comprehensive information across the field that includes graduate and undergraduate courses. Echoing the concerns of Keene (1997), I have to wonder, however, what is truly the difference between the two levels of courses. Keene (1997) encouraged the field to "pay attention to the differences among our levels of instruction" (p. 195), but, from course descriptions to course materials, there was no specific distinction in course outcomes between the undergraduate courses and the graduate courses. For example, the crossover courses provide evidence for this concern. For those institutions, the only difference in the courses is that graduate students are asked to do an additional assignment, lead discussions, or, in the words of one faculty member, "they are graded at a higher level" (from a Research 2 institution). Crossover courses and their ubiquity in TPC programs seem to underscore that there is definitely a grey area in need of exploration around the distinctions between a graduate-level and undergraduate-level editing course. The materials from the crossover courses make few—if any—distinctions between the student outcomes.

Having taught these types of courses myself and having talked with faculty about them, it seems the difference is played out in the level of assessment of student work. But even that distinction is muddied when things such as editing tests or exercises around style and copyediting are assessed in the same way for undergraduates as they are for graduate students.

Part of this could be attributed to the fact that many master's-level institutions recruit students on the premise that they do not need prior educational experience in TPC; thus, a course like editing would need to be designed assuming no prior knowledge. The lack of distinction or difference between the levels of the course raises broader programmatic questions about what it is we value and how we assess and describe that value and the differences in these courses to stakeholders outside of our programs. But if this is indeed the case, which is most likely, then it does mean TPC needs to have hard conversations about the broader aims and goals of a master's degree, including expectations for different types of masters' degrees (e.g., ones that are specifically geared to serve as something akin to a terminal professional degree like an MBA or those that are more focused on preparing students for a PhD program).

## Important Topics with Little Coverage

From the course descriptions to the topics within the syllabi there were some noticeable omissions: ethics, visual and/or design, and intercultural/global concerns. What is noticeably absent from any of these descriptions is a mention of or emphasis on ethics in the practice of editing. This was a noticeable omission since ethics only appeared in three of the course descriptions. Only four syllabi had weekly topic units on ethics. While ethics can definitely be embedded within class discussions, it is discouraging that there is not more of a direct emphasis on the importance of ethics, particularly situational ethics that will likely come up a number of times for most of our students during the course of their careers.

It was also noticeable and surprising that the editing course descriptions also did not regularly and consistently talk about the editing of visual or design aspects. There were only 12 course descriptions that explicitly discussed the visual: four references to visuals; two to document design; four to layout and design; and two to design principles. The course descriptions aligned with the syllabi, since there were 13 syllabi having a *single* week devoted to the topic. Some might counter this information with the idea that visual rhetoric or document design is integrated into other aspects of the course. This may be true at some locations; outside of managing fonts or headings with a style guide, there was little to no discussion in the course materials on how to edit visuals, information, or the document's design. The ongoing emphasis on information design and electronic delivery means that TPC programs are missing a huge opportunity (as the most generous way of saying it) to integrate and emphasize the need to better design information for current expectations in the workplace.

Finally, a glaring omission in the topics covered was the limited or nonexistent emphasis on editing for global or cultural contexts. While St.Amant's chapter in this volume provides some important guidelines, I would offer a call to action to include a greater emphasis within editing courses since only four institutions (two undergraduate and two graduate) had global issues as a single weekly topic.

It is important to point out that these omissions are from my own personal understanding of workplace practices, having spent over 20 years as a practicing consultant. This means that someone else could examine these same materials and find areas that need additional coverage. For those that may say there is no way to cover every topic in every course, I would counter with the idea that, based on the course materials I received, TPC faculty could easily reduce the number of weeks spent on copy marks and grammar and include at least an introduction to other important topics such as the ones mentioned here.

## Issues of Labor

Time constraints of getting this chapter completed prohibited a more thorough understanding of the labor issues involved in this course, but it is important that TPC always be aware of the teaching labor behind our curricular initiatives

(Melonçon, 2014, 2018). As an almost singular voice about labor issues in TPC (Melonçon, 2017; Melonçon & England, 2011; Melonçon, England, & Ilyasova, 2016; Melonçon, Mechenbier, & Wilson, forthcoming), I insist that the field cannot discuss curricula in any capacity without taking a moment to talk about labor issues and to review who is teaching TPC courses.

Unlike some of the other courses in the TPC curricula (e.g., the service course that is primarily taught by contingent faculty, see Melonçon & England, 2011), more tenure-line faculty teach the editing course. Of the 48 institutions (out of 86) where I identified the specific faculty who had taught the undergraduate editing course recently, 65% of those courses are being taught by tenure-line faculty. However, that still means that contingent faculty teach 35% of editing courses at the undergraduate level. Since I only examined institutions where editing is a required course, this fact means that the curricular mission of many departments could not be met without the reliance on contingent faculty. It is important to note that while labor issues are usually confined to undergraduate curricula, they can also impact graduate-level curricula. Half of the crossover courses and two of the graduate-level editing courses are taught by contingent faculty. Granted, the original reason that higher education justified adjunct labor was to find specialists to teach certain types of courses. This is exactly the case for the two graduate-level editing courses, but it is important to talk about all aspects of contingent labor when we are talking about TPC programs and staffing required courses.

Also, with the large number of editing courses offered, it is likely that there are more contingent faculty teaching than I have captured here. For example, in some cases, I could not find who taught the course, which means more than likely it is being taught by a term adjunct or a graduate student with some workplace experience. Or, the issue is obscured because I knew a tenure-line faculty who had taught the course and for ease of gathering information I stopped searching when I came across someone that I knew teaching the course.

While Lang and Palmer (2017) point out that "faculty may possess neither the comfort nor expertise with moving a technical editing class toward multimodality" (p. 307), I would counter that this may not necessarily be true. All of the tenure-line faculty (at the graduate level) have experience with teaching in different modes, as do many of those teaching at the undergraduate level. I can say this with confidence knowing the faculty, their scholarship, and their orientations to the field itself. It seems that the faculty issue in question is not about expertise or qualifications, but rather, simply the need for more tenure-line or long-term full-time faculty, and as the next section considers, what faculty and programs mean when they talk about editing courses.

## So What Is Editing? What Should It Be?

Across all degree types, around 85% of all institutions have an editing course, which makes it the most common course across all curricula in the United

States (outside of the service course, which is usually used for degree programs). Because of this fact, it seems TPC PAs and faculty have tacitly agreed that editing is an important skill for students to learn. However, the number of institutions that require that course is relatively low compared with the number of institutions where it is an option as part of the degree program. This raises the question, why is the course not required at more institutions? In other words, why does the field seem to agree that it is an important course, but yet does not require it?

The answer to that question seems to be keyed to a number of practical concerns. For example, a TPC PA from an R1 institution (highest research activity) explained that the editing course went from being required to being an elective because of the concerns over staffing. A required course must be taught regularly and this particular institution was facing faculty reduction issues (without replacement of tenure lines). While the course is still offered semiregularly, it is no longer required. In a separate example, the TPC PA from an R3 institution explained that while they feel editing is an important skill, the course wasn't required because editing is both implicitly and explicitly embedded in the majority of courses in the curriculum.

Meanwhile, editing still remains a vital component of the work of technical and professional communication. As one faculty member, who teaches editing at an R1 school, pointed out,

> Teaching rhetoric in the abstract is not really all that difficult, and basic editing skills can be taught to just about anyone who really wants to learn how to write clearly. Applying editing techniques with a rhetorical sensibility is the more difficult thing, but it is our way in to providing our students with applied rhetorical skills that really sets them apart from other professionals in the workplace.

It remains essential that, as a field, we have conversations about what an editing course is and whether the traditional focus on fundamental editing needs to shift, and, if so, how that can be done within the constraints of current curricula.

The tension of what is editing is even found in the book many institutions use, *Technical Editing*. In this textbook, Rude and Eaton (2011) posit the idea that editing is more than "grammar janitors," but the text itself still focuses most predominately on the editing fundamentals to the detriment of what they claim to be doing. For example, Rude explains that the "comprehensive editing process requires the editor to analyze the document's purpose, readers, and uses" and this "precedes editing for grammar, punctuation, and mechanics" (2005, p. 231). Yet, the book is organized so that the mechanics precedes the information on comprehensive editing. And, unfortunately, all the syllabi and materials I reviewed followed this pattern, which seems fundamentally backward to what TPC professes to be doing.

The tensions between Kreth and Bowen (2017) and Lang and Palmer (2017) exemplify the tensions within academic programs and the wide variety of approaches to the editing course. While, across the board, all editing courses provide students with the foundations of editing (the actual working with words and sentences and grammar, etc.), after this common trait the "editing" courses have little in common, even though the course descriptions and titles seem to indicate that they do.

Lang and Palmer (2017) advocate for multiple editing courses in the TPC curricula that include a "fundamentals" course, which would be the majority of courses currently being taught, followed by more different approaches to editing. The problem with Lang and Palmer's position is that there is little room in the curricula for their advanced courses, and as Kreth and Bowen's (2017) work illustrates, there is little demand for the multimodal skills Lang and Palmer describe.

"Rather than shrinking in importance and relevance, the work of technical editors continues to find new and diverse homes where technical editors' skills and strengths pay dividends" (Kreth & Bowen, 2017, p. 254). The prevalence of editing-type jobs in the workplace means that TPC needs to consider what it actually means by editing, particularly in light of the idea that many communication-type jobs—in which the majority of our students work—combine the author and editing role. Thus, editing is still a vital and important skill for programs to be teaching.

Programmatically, faculty and TPC PAs need to determine what their program means by editing and then how to best address that within their curricula. This could mean keeping and sustaining the current editing course with its focus on fundamentals, or it could mean expanding the course to include some of the aspects brought up by Lang and Palmer (2017), or it could mean not requiring the course and including components of it in other courses. All three of these models are currently being used in programs and courses across the United States.

It is clear that there is not one "editing," and, for TPC, we need to advance discussions around what editing does mean and, more importantly, what definition will guide the creation of "editing" courses. Having spent a considerable amount of time over the last several years examining programs and courses, and talking to faculty and administrators, I have learned that there is never any easy one-size-fits-all answer to any curricular question. But there are always better answers than others, and the better answers always rely on having data available from across the field to match them to local circumstances. Therefore, the editing course does not need a major overhaul. It needs a better definition of what it is with a visible declaration of that definition so that students and stakeholders know how that program defines the course. The course should include editing fundamentals and information on how editing works within current workplaces in a more distributed and/or collaborative fashion. There must be an emphasis on visuals and design, ethics, and intercultural concerns. Courses need a greater emphasis on comprehensive editing (which by its very nature includes fundamental editing) that provides students with the opportunities to practice the full range of editing and what it means to do that work, from line editing, to working with difficult collaborators, to managing the process through technologies. And as important, this practice is situated within a specific context that ensures students are getting additional exposure to the complexities of purpose and audience.

When asking what editing is, I am also encouraging TPC PAs and faculty to consider where these other types of skills—those mentioned by Lang and Palmer (2017) and also those mentioned in a variety of published literature (see Henschel & Melonçon, 2014, table 2, pp. 11–12; c.f., Lanier, 2017)—are currently being taught. Programmatically, some of what Lang and Palmer attribute to an editing course is actually already going on in other parts of the TPC curricula, such as in technology-centric courses like information design and content management.

One of the reasons I have advocated for moving programmatic research to the field-wide level (Melonçon, 2018) is because often faculty make moves to curricula based on their limited view of what the field is and what the field needs. The recent move to examine job ads as a way to calibrate programs (see, e.g., Brumberger & Lauer, 2017; Stanton, 2017) is misguided because it fails to place that information into both a local and field-wide academic context. It also fails to recognize the limitations of the job ad as an artifact in determining what is actually happening in the workplace. For TPC PAs and faculty, asking the questions "what is editing" and "what does it add to a program" are vital starting places in determining how singular courses fit into an overall vision in preparing students for the diverse range of jobs they take on in a diverse range of industries and professional settings.

## Pedagogical Applications

The data and implications provided here should provide instructors with insights into current trends across the field in the United States, about what editing is and what it is doing. This information can assist faculty and program administrators with updating their courses or, at the very least, with having a conversation about the role of the editing course in the curriculum and what the course should be doing for students.

## Looking Ahead

The editing course remains the most prominent "core course" across the field, but until now—some 60 years after the field's first degrees—TPC did not have a thorough understanding of the course and the pedagogical approaches to it.

The day-to-day life of faculty and TPC PAs often mean being embedded in local institutional cultures to the point that we need to be reminded of the larger field and what can be referred to as effective practices and approaches. Overviews like this one are important to TPC because they afford the field the opportunity to see things at the macrolevel (field-wide level) and microlevel (course level).

Understanding the multiple layers of editing from the document's purpose and audience to issues of consistency in visual design to sentence level copyediting are all essential skills that any practicing technical and professional communicator needs to know. The editing course remains a vital and key component to TPC curricula. However, how editing skills—from fundamentals to more advanced issues in different modes and mediums—are integrated into the TPC curricula is one that needs more intensive scrutiny, debate, and research.

## Pedagogical Practicalities

The information presented can be used by TPC PAs and program faculty to determine how their local courses match, or not, to current trends across the United States. Faculty and administrators should use this field-wide information as a way to place their local course and localized contexts, that is, how their students are using these skills in their jobs, to update and to keep their editing courses relevant and useful for students.

## Acknowledgments

Research such as this is not possible without the generosity and trust of colleagues in the field. I want to send out a big thank you to the faculty who made this work possible by sending their course materials to me and by answering questions about their courses. I remain so grateful for their help. However, I maintain sole responsibility for the analysis and interpretation of those materials.

## Note

1 This project was completed under Institutional Review Boards at University of Cincinnati #2012-4082 and University of South Florida #Pro00033052. The oddities of dual IRBs due to my changing institutions means that I am unable to include a list of institutions included in this analysis. However, I can provide interested readers with the full list of editing courses from TechComm Programmatic Central.

## References

Boettger, R. (2014). The technical communication editing test: Three studies on this assessment type. *Technical Communication, 61*(4), 215–231.

Brumberger, E., & Lauer, C. (2017). The evolution of technical communication: An analysis of industry job postings. *Technical Communication, 62*(4), 224–243.

Chong, F. (2016). The pedagogy of usability: An analysis of technical communication textbooks, anthologies, and course syllabi and descriptions. *Technical Communication Quarterly, 25*(1), 12–28. doi:10.1080/10572252.2016.1113073

Crovitz, D., & Devereaux, M. (2017). *Grammar to get things done: A practical guide for teachers anchored in real-world usage.* New York, NY: Routledge and National Council for Teachers of English.

Dubinsky, J. M. (2002). More than a knack: Techne & teaching technical communication. *Technical Communication Quarterly, 11*(2), 129–145. doi:10.1207/s15427625tcq1102_2

Henschel, S., & Melonçon, L. (2014). Of horsemen and layered literacies: Assessment instruments for aligning technical and professional communication undergraduate curricula with professional expectations. *Programmatic Perspectives, 6*(1), 3–26.

Keene, M. L. (Ed.) (1997). *Education in scientific and technical communication: Academic programs that work.* Arlington, VA: Society for Technical Communication Press.

Kreth, M., & Bowen, E. (2017). A descriptive survey of technical editors. *IEEE Transactions on Professional Communication, 60*(3), 238–255. doi:10.1109/TPC.2017.2702039

Lang, S., & Palmer, L. (2017). Reconceiving technical editing competencies for the 21st century: Reconciling employer needs with curricular mandates. *Technical Communication, 64*(4), 297–309.

Lanier, C. (2017). Toward understanding important workplace issues for technical communicators. *Technical Communciation, 65*(1), 66–84.

Melonçon, L. (2009). Master's programs in technical communication: A current overview. *Technical Communication, 56*(2), 137–148.

Melonçon, L. (2014). Curricular challenges of emphasis degrees in technical and professional communication. In T. Bridgeford, K. S. Kitalong, & B. Williamson (Eds), *Sharing our intellectual traces: Narrative reflections from administrators of professional, technical, and scientific, communication programs* (pp. 179–200). Amityville, NY: Baywood.

Melonçon, L. (2017). Contingent faculty, online writing instruction, and professional development in technical and professional communication. *Technical Communication Quarterly, 26*(3), 256–272. doi:10.1080/10572252.2017.1339489

Melonçon, L. (2018). Critical postscript: On the future of the service course in technical and professional communication. *Programmatic Perspectives, 10*(1), 201–230.

Melonçon, L., & England, P. (2011). The current status of contingent faculty in technical and professional communication. *College English, 73*(4), 396–408.

Melonçon, L., England, P., & Ilyasova, A. (2016). A portrait of non-tenure-track faculty in technical and professional communication. *Journal of Technical Writing and Communication, 46*(2), 206–235. doi:10.1177/0047281616633601

Melonçon, L., & Henschel, S. (2013). Current state of US undergraduate degree programs in technical and professional communication. *Technical Communication, 60*(1), 45–64.

Melonçon, L., Mechenbier, M., & Wilson, L. (forthcoming). A national snapshot of the material working conditions of contingent faculty in composition and technical and professional communication. *Academic Labor: Research and Artistry.*

Melonçon, L., & St. Amant, K. (2018). Empirical research in technical and professional communication: A 5-year examination of research methods and a call for research sustainability. *Journal of Technical Writing and Communication.* Advance online publication. doi:10.1177/0047281618764611

Melançon, L., & Schreiber, J. (2018). Advocating for sustainability: A report on and critique of the undergraduate capstone course in technical and professional communication. *Technical Communication Quarterly, 27*(4), 322–335. doi:10.1080/105 72252.2018.1515407

Rude, C. D. (2005). *Technical editing* (4th ed.). Allyn & Bacon Series in Technical Communication. New York, NY: Longman.

Rude, C. D., & Eaton, A. (2011). *Technical editing* (5th ed.). Boston, MA: Longman.

St.Amant, K., & Melançon, L. (2016). Reflections on research: Examining practitioner perspectives on the state of research in technical communication. *Technical Communication, 63*(4), 346–364.

Stanton, R. (2017). Do technical/professional writing (TPW) programs offer what students need for their start in the workplace? A comparison of requirements in program curricula and job ads in industry. *Technical Communication, 64*(3), 223–236.

Taylor, S. S. (2006). Assessment in client-based technical writing classes: Evolution of teacher and client standards. *Technical Communication Quarterly, 15*(2), 111–139. doi:10.1207/s15427625tcq1502_1

Williams, J. M. (1981) Phenomenology of error. *College Composition and Communication, 32*(2), 152–168.

Youngblood, S. A., & Mackiewicz, J. (2013). Lessons in service learning: Developing the service learning opportunities in technical communication (slot-c) database. *Technical Communication Quarterly, 22*(3), 260–283. doi:10.1080/10572252.2013.77 5542

# INDEX

accessibility 24
activities 22
agency 67
agency network 80
analysis 7
analytical thinking 50, 53, 56, 60
apparency 104
apparent feminist theory 104
artificial intelligence 15, 30
ASD Simplified Technical English 137-8
assessment 177-8, 182, 183; benchmarks 29; editing tests 25
assignments 172, 177
audience 6, 85, 92-3, 98, 104, 111-12, 122; analysis 85, 148-50; awareness 29; global 19, 30, 150
authenticity 50, 53, 58, 60
author-editor relationships 19, 25, 178
authors 92, 95
automation 142; macros 33

benchmarks 29

career as technical editor 19, 23-4, 36
client-based projects 180
clout 50, 53, 59
collaboration 24, 33, 36, 40, 82, 105, 123, 130, 135-6, 140-1, 178, 180
comments 19, 30, 112, 114, 119, 123, 125, 130; global 118
communication problems 18, 23-4, 36, 111, 125

communities of practice 69, 71
complex sentences 156-8
comprehensive editing 6, 31, 110-12, 124, 176, 178, 181, 186
content audit 82
content management system 28, 35, 67
content model 87
controlled language 41, 138, 164-5
convenience 133
copyediting 5, 22, 181
course descriptions 172, 175, 184, 187
course titles 171, 174
crossover course 172, 174, 183
culture 30-1, 152, 184; workplace 34, 36
curricula 25, 36-7

definition, editing 17, 20-2, 185-7
definition, technical editing 17, 20-2, 185-7
degree: graduate 171, 174; undergraduate 172-3
degree program 172-3
delivery 93
developmental editing 22, 32, 125
discrimination 95
DITA 142
dynamic content 35, 81, 87

economies of practice 70
editing: comprehensive 6, 31, 110-12, 124, 176, 178, 181, 186; definition 17, 20-2, 185-7; developmental 22, 32, 125;

substantive *see* comprehensive editing; technical *see* technical editing; vs. revision 22
editing courses 171–89
editing fundamentals 180
editing internships 36
editing of content: dynamic content 35, 81, 87; priorities 40; triage approach 23; visuals 184
editing priorities 40
editing projects 178–9
editing tests 25
editorial analysis 7
editorial comments 19, 30, 112, 114, 119, 123, 125, 130
electronic editing 32, 123
emotion 102
emotional tone 50, 53, 55
empathy 102
empirical research 2, 23
enculturation 36, 40
English, global 146, 150, 154, 166, 184
equality 95, 98, 100, 106
errors 25, 31, 112
ethics 20, 26, 184

feminism 9, 94
feminist theory 91
freelance technical editor 36
fundamental skills 19, 180, 188

gender 94, 103
gender imbalances 96
GitHub 136, 140
global comments 118
global English 146, 150, 154, 166, 184
global level edit 111
glossary 162, 165, 167
Google Docs 82, 130, 135, 139, 141
graduate courses 174
graduate degree 171, 174
grammar 5, 19, 112, 182

harassment 95
hardcopy editing 33
hardcopy markup 33
history of technical editing 23, 92
human–information interaction 32, 40, 110
humanistic 67

identity of participation 67
idioms 158–9

imagination 69–70
information architecture 30, 35, 154
interaction, human–information 32, 40, 110
international 10
international audiences 19, 30, 150
internships 36
interpersonal skills 25, 40
intersectional feminist theory 94, 103

job titles 24, 27

knowledge transfer 5

labor 184–5
language 19; controlled 41, 138, 164–5; plain 19–20; second 19, 24; translation 41
LaTeX 136–7
learning transfer 5
levels of edit 17, 20, 22, 111, 125, 177

macros 33
MadCap Flare 130, 136, 138
markup 23, 30, 129; hardcopy 33; metadata 30, 35
metaphors 101, 159, 161
Microsoft Word 61, 130, 133, 138, 143; track changes 32–4, 52, 57, 60, 130, 133, 135

narrative accrual 76
narrative ways of knowing 67
noun strings 154–5, 157
numbers 160–1

Open Office 130
order-of-teaching 3
Overleaf 136
ownership of a meaning 80

parallelism 152
passive voice 154–5, 157
pedagogical approaches 171, 179, 183, 188
pedagogy, technical editing 35, 40–1, 61–2, 84–8, 104–7, 142–4, 166–8
peer review 22, 28, 134, 139
PerfectIt 141
personas 85
plain language 19–20
power structures 100
practice, economies of 70
program, degree 172–3

programmatic approach 92, 111
*Programmatic Perspectives* 75
project management 36
projects, client-based 180
pronouns 54, 59, 153
proofreading 22, 24

quality control 23, 27–8

regex *see* Regular Expressions
Regular Expressions 137, 141
respect 100
revision vs. editing 22
rhetoric 9, 21, 29, 179, 186
rhetorical theory 92
roles of technical editors 19, 27, 35–6, 187

second language 19, 24
sentiment analysis 50, 52
SharePoint 135
single sourcing 70
skills 19, 25, 40, 180, 188
Society for Technical Communication 18, 33, 130–1, 180
specialized editing 9
status of technical editors 27–8
STC *see* Society for Technical Communication
structured content 143
style 178
style guide 141, 178
style sheets 19, 26, 73, 86, 163, 167
StyleWriter 137
substantive editing *see* comprehensive editing

teaching methods 3
technical editing: activities 22; curricula 25, 36–7; definition 17, 20–2, 185–7; electronic editing 32, 123; empirical research 2, 23; hardcopy editing 33;

history 23, 92; pedagogy 35, 40–1, 61–2, 84–8, 104–7, 142–4, 166–8; skills 19, 180, 188; textbooks 35, 112, 123, 186; tools 24, 32, 34; triage approach 23
technical editors: career 19, 23–4, 36; freelance 36; interpersonal skills 25, 40; job titles 24, 27; roles 19, 27, 35–6, 187; status 27–8; value 19, 24, 27–8, 186
technology 11, 19, 32–3, 35, 40, 129–30
TechScribe 137
text, tracked changes 32–4, 52, 57, 60, 130, 133, 135
textbooks 35, 112, 123, 186
theory 26, 34; apparent feminist 104; feminist 91; intersectional feminist 94, 103; rhetorical 92
tools 24, 32, 34
topic-based writing 69, 86–7
track changes 32–4, 52, 57, 60, 130, 133, 135
translation 41, 138, 147, 149, 162
triage approach 23

undergraduate degree 172–3
usability 23–4, 148
user experience 29
user-generated content 31, 35

validation 137
value of technical editors 19, 24, 27–8, 186
version control 129, 135
visuals, editing of 184
voice 123–4

ways of knowing 67
workplace culture 34, 36
workplace practices 180
writing, topic-based 69, 86–7

XML 86